Western Region Renewable Energy Markets: Implications for the Bureau of Land Management

Scott Haase, Lynn Billman, and Rachel Gelman

Prepared under Task No. WFH7.1004

NREL is a national laboratory of the U.S. Department of Energy, Office of Energy Efficiency & Renewable Energy, operated by the Alliance for Sustainable Energy, LLC.

National Renewable Energy Laboratory
1617 Cole Boulevard
Golden, Colorado 80401
303-275-3000 • www.nrel.gov

Technical Report
NREL/TP-6A20-53540
January 2012

Contract No. DE-AC36-08GO28308

NOTICE

Preface

The purpose of this analysis is to provide the U.S. Department of the Interior (DOI) and the Bureau of Land Management (BLM) with an overview of renewable energy (RE)[1] generation markets, transmission planning efforts, and the ongoing role of the BLM RE projects in the electricity markets of the 11 states (Arizona, California, Colorado, Idaho, Montana, Nevada, New Mexico, Oregon, Utah, Washington, and Wyoming) that comprise the Western Electricity Coordinating Council (WECC) Region. Specific objectives are to:

- Determine current levels of renewable energy generation in the WECC

- Provide an estimate of future demand based on requirements arising only from state renewable portfolio standards (RPS) existing as of June 2011. For purposes of this analysis, we assume there will be no new national or state-level legislation for clean energy policies

- Develop an estimate of renewable generation that is under development[2] in the region

- Estimate the current balance between planned supply (existing and projected projects) and projected demand from RPS requirements for renewable energy

- Assess how the planned renewable energy projects of DOI and the BLM fit within this broader market perspective

- Provide an update on regional transmission planning efforts and assess the impact of these efforts on renewable energy development in general and the BLM renewable energy program in particular

- Develop suggestions for increasing the strategic value of BLM projects and synergies across all federal lands.

This analysis focuses on the status of, and projections for, likely development of non-hydroelectric renewable electricity from solar (including photovoltaic [PV] and concentrating solar power [CSP]), wind, biomass and geothermal resources in these states.

The study uses recent data available from the BLM, U.S. Energy Information Administration (EIA), WECC, Lawrence Berkeley National Laboratory (LBNL), state public utility commissions, state energy offices and SNL Financial LC (SNL). A detailed list of sources is provided in the References section.

---

[1] Throughout this report, the term "renewable energy" refers to electricity generated from renewable sources such as wind, solar, geothermal or biomass.

[2] The focus is limited to projects that are under construction, or otherwise under advanced stages of development. "Advanced Development," using the SNL Financial definition, includes projects that meet two of the following criteria: have a signed power purchase agreement, obtained financing, procured turbines or other major parts, hired an engineering, procurement and construction (EPC) contractor, or obtained all permits.

Caveats

Several important caveats should be noted for this analysis.

First, estimates of RPS requirements are subject to uncertainty in future retail sales because RPS percentage targets are applied to retail sales. Market conditions change frequently; thus, key parameters affecting the results of this analysis will be fluid and change frequently. A case in point is Xcel Energy's recent filing with the Colorado Public Utilities Commission (made too late for incorporation into this analysis) of its most recent 10-year resource plan (Xcel 2011). In the new plan, Xcel states that it expects to require significantly less capacity by 2018, because of slower economic growth and other factors such as the success of its rooftop solar and demand-side management programs.

Second, this analysis is partly based on planned project data that reflect a "snapshot in time." Planned project estimates are based on the best available planned project data compiled from June through late November 2011 and summarize projects that are either under construction or under advanced development. In all likelihood, some planned projects will change in scope or size; some will be delayed; some will never be built; and other unexpected projects will emerge. Therefore, all quantitative projections in this analysis must be considered as estimates that indicate possible trends.

Third, the quantitative analysis is based on RPS-driven projections. However, other market drivers are also important in shaping this evolving market, including other federal and state incentives, utilities' decisions driven by evolving renewable energy economics and transmission availability, and the desire of some states to diversify their economies and create employment. Projections of demand for these types of market drivers are not readily available. Therefore, until analyses are available for other market drivers, all quantitative projections in this report must be considered as estimates that rely primarily on known policy goals under current state level RPSs.

Finally, capacity-based analysis of the supply-demand gap is based on an average capacity factor across all technologies. Most state RPS requirements are based on energy requirements; therefore the capacity requirements to meet state RPS will vary depending upon the specific mix of renewables installed.

Acknowledgments

This work was funded by the Bureau of Land Management. The authors wish to thank Steve Black from the Department of the Interior for his early vision to conduct this analysis, and Ray Brady of the Bureau of Land Management for providing funding and access to data from the BLM's renewable energy program. The authors also wish to thank Galen Barbose of Lawrence Berkeley National Laboratory for his timely review of the draft paper as well as useful suggestions on the analytic approach and interpretation of LBNL's data. Jim Baak (Vote Solar) and Paul Tigan (BLM) provided external review and useful comments on the paper. Finally, the authors would like to thank NREL staff who provided review and input including Deborah Sandor, Jenny Heeter, David Hurlbut, Dylan Hettinger, Lori Bird, Dan Billelo, David Kline, Gian Porro, and Scott Gossett, as well as NREL Communications for editorial and publishing support.

List of Acronyms

BLM   Bureau of Land Management
CSP   concentrating solar power
DOI   U.S. Department of Interior
DRECP   Desert Renewable Energy Conservation Plan
EIA   U.S. Energy Information Administration
EPAct   Energy Policy Act of 2005
GIS   geographic information system
GWh   gigawatt-hour
LBNL   Lawrence Berkeley National Laboratory
MW   megawatt
NREL   National Renewable Energy Laboratory
PEIS   Programmatic Environmental Impact Statement
PV   photovoltaic
RE   renewable energy
REC   renewable energy certificate
RFDS   reasonably foreseeable development scenario
RPS   renewable portfolio standard
RTEP   regional transmission expansion planning
SEZ   solar energy zones
SNL   SNL Financial LC
WECC   Western Electricity Coordinating Council
WEPA   wind energy priority areas
WGA   Western Governors' Association
WREZ   western renewable energy zones

Executive Summary

At the request of the U.S. Department of the Interior (DOI) and the Bureau of Land Management (BLM), the National Renewable Energy Laboratory conducted an analysis to estimate the gap between the renewable energy (RE) generation supply and demand in the western states, and how that impacts BLM activities in this sector. The purpose of this analysis is also to provide DOI and BLM with an overview of RE markets, transmission planning efforts, and the role of BLM RE projects in the electricity markets of the 11 states (Arizona, California, Colorado, Idaho, Montana, Nevada, New Mexico, Oregon, Utah, Washington, and Wyoming) that comprise the Western Electricity Coordinating Council (WECC) Region.

Without new policies at either the state or national level, and without the extension of special federal programs[3] that support the development of renewable electricity facilities, current state renewable portfolio standards (RPS) requirements will remain the primary driver for new RE deployment in the western United States. The quantitative portion of this analysis is based on RPS-driven demand, because projections for that demand are available. Other market drivers will also play a part in motivating investment decisions, such as the voluntary consumer market, other federal and state incentives of various types, utility decisions driven by improving RE economics and transmission availability, and other policies that arise from the desire of some states to diversify their economies and create green jobs. Modeling the impacts of these other drivers was beyond the scope of this analysis.

In total, the WECC states in 2010 generated more than their total RPS requirement for RE [53,500 gigawatt-hours (GWh) generated compared to 49,500 GWh required for RPS policies]. To analyze the possible gap between demand and supply in RE in 2020, three datasets were compared: one for demand and two for supply. The comparison was made in terms of capacity, rather than generation, to meet the needs of the BLM.

- The source data for future demand was taken from analysis by Lawrence Berkeley National Laboratory (LBNL) of state RPS requirements (Barbose 2011), and includes projected RE generation required in 2020 to meet today's RPS requirements. The total RPS-required demand for WECC states in 2020 is estimated to be 134,000 GWh. To convert this generation requirement into a demand in terms of capacity, three different approaches to capacity factors were used: LBNL's approach based on an average capacity factor of 50%, a scenario using a lower capacity factor (26%), and a scenario using a higher capacity factor (53%). These different approaches estimated that WECC states in 2020 may require installed capacity in the range of 28,000 MW to 46,000 MW, as shown in Table ES-1. This required capacity does not include RE demand that may result from voluntary consumer demand or other types of market drivers. Various market factors could increase or decrease these market estimates. States could revise RPSs downward or electricity demand could decrease in the future. As a point of reference, nationwide voluntary consumer demand markets totaled 35,000 GWh in 2010.

---

[3] Such as the Department of Energy Loan Guarantee Program, or the Treasury Department's "Payments for Specified Energy Property in Lieu of Tax Credits: Section 1603 Grants" program.

- One source of data for supply was the current (2010) installed capacity of RE in WECC states. Actual installed nameplate capacity in December 2011 was about 18,400 MW (SNL 2011b), as shown in Table ES-1.

- The other source of data for supply was taken from planned projects as tracked by SNL Financial (SNL 2011a). Only planned projects that are under construction or in advanced development[4] were included in this analysis; the planned capacity of these projects totals about 18,900 MW. The output from planned projects was assigned to states based on the location of the power purchaser, if known (about two-thirds of the planned capacity); if not known, the output was assigned to the state where it is planned to be located (about one-third). Some larger power purchasers sell their power across several states. In those cases, a recent analysis (Hurlbut 2011) examining each utility's power purchase agreements was used to apportion the output. The sum of actual installed capacity today and planned capacity is about 37,000 MW, as shown in Table ES-1.

- Compared to the range of required RE capacity, Table ES-1 shows that the gap between demand and supply in 2020 ranges from a potential oversupply of 8,000 MW to an unmet demand of 9,000 MW. Because the demand estimates were only based on meeting RPS requirements, other types of market drivers, as mentioned above, should also be investigated within each state, to present a more complete picture of 2020 demand.

---

[4] "Advanced Development," using the SNL Financial definition, includes projects that meet two of the following criteria: have a signed power purchase agreement, obtained financing, procured turbines or other major parts, hired an engineering, procurement and construction (EPC) contractor, or obtained all permits.

Table ES-1. Renewable Energy 2020 Supply/Demand Summary Based on RPS Requirements

			Arizona	California	Colorado	Idaho	Montana	Nevada	New Mexico	Oregon	Utah	Washington	Wyoming	Total WECC
SUPPLY	As of December 2011 – Not Including Hydropower[4]		Existing Renewable Energy Plant Capacity (MW)[1] (SNL 2011c)	194	7,880	1,372	441	404	589	738	2,558	287	2,547	1,419	**18,429**
			Planned Renewable Energy Projects Capacity (MW) (SNL 2011a)	1,045	10,367	91	564	300	1,374	130	864	0	1,245	2,841	**18,820**
			Sum of Existing and Planned Renewable Energy Capacity (MW)	1,239	18,247	1,462	1,005	704	1,963	868	3,422	287	3,792	4,260	**37,249**
DEMAND	Capacity Required in 2020 to Meet Current RPS Requirements (MW)[2]	A	High Capacity Factor, Low Capacity (MW)	1,380	18,653	2,741	0	325	958	896	1,733	0	1,807	0	**28,493**
		B	LBNL Capacity Factors (Barbose 2011) (MW)	1,415	27,944	2,741	0	382	896	985	1,811	0	2,952	0	**39,126**
		C	Low Capacity Factor, High Capacity (MW)	2,144	31,354	2,741	0	662	1,421	896	3,079	0	3,238	0	**45,535**
GAP	Oversupply or (Unmet Demand) in 2020[3]	A	High Capacity Factor, Low Capacity (MW)	(141)	(406)	(1,279)	1.005	379	1,005	(28)	1,689	287	1,985	4,260	**8,756**
		B	LBNL Capacity Factors (Barbose 2011) (MW)	(176)	(9,697)	(1,279)	1,005	322	1,067	(117)	1,611	287	840	4,260	**(1,877)**
		C	Low Capacity Factor, High Capacity (MW)	(905)	(13,107)	(1,279)	1,005	42	542	(28)	343	287	554	4,260	**(8,286)**

Notes:
1. These data are from Table 1 (SNL 2011c), and do not distinguish between projects responding to RPS and projects responding to other market drivers.
2. See text for description of the three capacity factor scenarios.
3. Various market factors could increase or decrease these market estimates.
4. Some of the 3,833 MW of planned hydropower will likely qualify for an RPS; however, RPS regulations on hydropower are complicated, and for simplicity, hydropower was omitted from this estimate.

The BLM is required by the Energy Policy Act of 2005 to approve 10,000 MW of non-hydropower renewable energy to be located on public lands by 2015. To date, the BLM has approved or authorized more than 5,200 MW of RE on federal lands.[5] Another approximately 8,000 MW in advanced development is included in the BLM list of 2011 and 2012 high priority projects. Based on this list and the progress made in 2011, the BLM appears to be on track to meet its EPAct goal of authorizing 10,000 MW by 2015, and indeed is likely to accomplish this goal by late 2012 or early 2013. **The 5,200 MW of approved or authorized BLM projects represent 11%-18% of the projected capacity additions needed to meet RPS requirements in WECC states by 2020**.

It should be noted in Table ES-1 that California is the primary driver for RE development across the WECC, representing over half of projected 2020 demand. California's demand has in-state as well as regional implications for transmission, and its policy environment moving forward will be a critical influence on future RE supply/demand balance.

The transmission portion of the analysis concludes:

---

[5] After analysis for this paper was completed, BLM announced their approval of an additional 1,400 MW of projects, bringing the total approved to 6,600 MW between December 2009 and December 2011.

- In total, WECC's highest priority ("foundational/common case") transmission projects would add more than 5,500 line miles, which may ultimately improve the viability of some of BLM's planned projects (WECC 2011).

- The interagency Rapid Response Transmission Team, initiated in 2011 through the U.S. Council on Environmental Quality, recently announced a coordinated and expedited permitting process for seven pilot transmission lines. Five of these lines are located in the western United States, and the BLM is the lead agency on four of these five lines. The focus on these lines represents a concentrated effort to connect potential renewable energy supplies with loads in the northwest and in southern California.

- Most of the BLM's priority wind and geothermal projects are within 20 miles of existing transmission lines; most of BLM's priority solar projects are within five miles of existing transmission lines. The analysis of whether there is capacity available on these lines, or whether they could be upgraded, has not been undertaken at this time.

- In general, new RE development projects sited close to load centers are less likely to be impacted by constraints of the current transmission infrastructure over the next 10 years; however, for more remotely sited RE projects, which will likely include some of the BLM projects, additional transmission infrastructure will be required. The additional 19,600 miles of new transmission lines (between 115 and 500 kV) currently planned for WECC states will support some of the expansion required for RE deployment required to meet western state mandatory RPS requirements. In some cases, the cost of some renewable resources located remote to load may offer the potential to reduce overall costs to ratepayers of meeting RPS requirements (WECC 2011).

Suggestions for the BLM Renewable Energy Program

Based on the analyses of supply and demand in WECC states, and BLM's interests in leasing land for RE projects, a number of specific suggestions are provided below to advance BLM efforts in this sector.

Update the RE project list. The information on BLM projects presented in this report changes frequently. Reviewing and updating the status of the projects on the master RE project list will help BLM prioritize necessary actions. The BLM Washington office has recently issued a call for information and data for geographic information systems (GIS) analysis of the wind and solar projects that the BLM's state and field offices are tracking. This information will be useful for updating the status of projects for FY12.

Focus on high-value project sites. In addition to the approved and high priority projects, BLM has received applications for more than 58,000 MW of renewable energy projects on public lands. The integration of BLM RE projects with planned transmission lines (especially the five pilot lines) will take on greater significance over the next few years, and BLM lands within potential interconnection distance of these lines are likely to see increased interest by industry, provided that these lands also meet all other suitability criteria for development. Further, any suitable BLM lands that are located close to load centers in states that are falling short of RPS requirements, or are located in regions that can potentially export to California, may see increased interest from developers. BLM should consider screening these lands, the solar energy zones, and other regions undergoing landscape-level planning (e.g., the Desert Renewable Energy Conservation Plan [DRECP], the West Chocolate Mountains Renewable Energy Evaluation Area, and the Arizona Restoration Design Project), against criteria designed to identify prime sites for future competitive leasing requests.

Work with other federal agencies and developers to facilitate project siting. A number of federal agencies, including other DOI bureaus and agencies, the Department of Defense, Department of Homeland Security, Department of Agriculture, Department of Energy, and Department of Commerce are interested in deploying renewable energy technologies to meet their internal mission goals. As specific examples, the Environmental Protection Agency's RE-Powering America's Land program seeks to promote the development of renewable energy projects on brownfield sites such as abandoned mining lands, landfills, and contaminated lands. The Bureau of Indian Affairs and many tribes are working to develop renewables on tribal lands. Developers are also targeting state, local, and private lands. Similar to the approach being taken by the BLM's Restoration Design project in Arizona, BLM could benefit from continuing to work cooperatively with other agencies and industry to identify and better understand the optimal suitable locations for development, regardless of land ownership.

Site projects to help support critical national needs. Similar to the strategy of siting RE to take advantage of drivers and interests of other agencies and private developers, there may be opportunities to increase the strategic value of BLM's RE projects by co-locating in prime areas that would also support national or regional energy security and resiliency, and support national environmental goals. As an example, BLM projects could be sited in locations that, in an emergency situation, could help supply power for critical loads such as water pumping and treatment facilities, hospitals, military installations, National Guard facilities, critical substations, radar sites, data centers, and other high value loads. In some cases, BLM may choose to work with the developers and recommend a shift of the BLM projects to other locations in the region that may offer greater strategic advantages. Continuing to avoid projects on environmentally sensitive lands will support national environmental goals. GIS and additional landscape level planning analyses such as those being undertaken by the DRECP and WECC's Environmental Data Task Force can help identify these specific locations and opportunities. Specific examples and case studies could be assessed in greater detail.

Identify options to integrate projects into existing fossil fuel generation. Siting RE projects near old, retiring or seldom used fossil fuel plants takes advantage of existing infrastructure and potential synergies. For example, BLM lands located near existing coal or gas plants may be candidate sites for solar thermal plants that are constructed from the outset to integrate fully into existing facilities. The National Renewable Energy Laboratory (NREL) and the Electric Power Research Institute recently completed an initial screening of fossil fuel plants that may be suitable for solar augmentation (NREL 2011). However, additional detailed work is needed in this area.

Table of Contents

Preface iii

Acknowledgments v

List of Acronyms vi

Executive Summary vii

    Suggestions for the BLM Renewable Energy Program x

List of Figures xiii

List of Tables xiv

Background 1

    Current Renewable Energy Markets in the Western United States 1

        Total Markets 1

        RPS-Driven (Compliance) Markets 5

        Voluntary (Consumer-Driven) Markets 6

Projections for Estimating the Supply-Demand Gap in the Western Region 8

    Demand Projections, 2010-2035 8

    Supply: Planned Projects by 2020 11

    Demand: RPS Requirements in 2020 12

Projected Gap in RE Capacity in 2020 14

BLM Renewable Energy Activities 15

    Programmatic Environmental Impact Statements 16

    The BLM's Reasonably Foreseeable Development Scenario – Solar PEIS 17

    Supplement to the Draft Solar PEIS – Transmission Analysis 18

    BLM's Priority Projects 18

    Geographic Locations of BLM Projects 22

        BLM Solar Projects 24

        BLM Wind Projects 25

        BLM Geothermal Projects 26

Regional Transmission Planning Status 27

    Western Renewable Energy Zones and EPAct Corridors 28

Strategic Opportunities for BLM Renewable Energy Projects 30

    Alignment with Renewable Energy Zones 31

    Alignment with Existing and Planned Transmission Expansion 31

    Coal Plant Retirements 35

    Co-location of BLM Projects with Fossil Fuel Plants 36

Conclusions 37

References 39

Appendix A 42

Appendix B 43

List of Figures

Figure 1: Total (all fuels) state electricity consumption and generation in 2010. 2

Figure 2. WECC generation mix in 2010 3

Figure 3. Estimated state electricity generation by fuel source for 2010..................................... 4

Figure 4. Growth in national voluntary green power market segments, 2006-2010. 7

Figure 5. Voluntary utility green pricing and competitive market power sales by state in 2010... 8

Figure 6. Projected electricity demand by state. 9

Figure 7. Projected RPS-applicable demand by state – WECC states.. 10

Figure 8. Projected total renewable energy generation required to meet RPS – WECC states.... 11

Figure 9. Overview of BLM projects proposed for the western United States.............................. 23

Figure 10. Overview of BLM solar projects proposed for the western United States.................. 24

Figure 11. Overview of BLM wind projects proposed for the western United States.................. 25

Figure 12. Overview of BLM geothermal projects proposed for the western United States........ 26

Figure 13. EPAct 368 corridors in the western states (Nov 2008). 30

Figure 14. BLM projects within five miles of current, foundational, or BLM lines. 32

List of Tables

Table ES-1. Renewable Energy 2020 Supply/Demand Summary Based on RPS Requirements . ix

Table 1.   State Renewable Nameplate Capacity (MW) as of December 2011 5

Table 2.   Renewable Energy Generation Required for Mandatory State RPS in 2010 Compared to Actual Renewable Energy Generation in 2010 Excluding Hydropower 6

Table 3.   Estimated Capacity of Currently Planned Renewable Energy Projects in Western Region 12

Table 4.   Calculations to Arrive at Lowest (26%) and Highest (53%) Capacity Factors for Use in 2020 Estimates (Tables 5 and 6) 13

Table 5.   Estimation of Capacity (MW) Based on Generation (GWh) Needed to Meet RPS Targets in 2020 14

Table 6.   Projected Oversupply or Unmet Demand for RE Capacity in 2020 in Western States 15

Table 7.   Six States: Comparison of Solar PEIS Projections for 2030 and Projections Based on RPS-Driven Demand in 2020 18

Table 8.   Planned Renewable Nameplate Capacity by State on BLM Land 20

Table 9.   Planned Renewable Generation by State on BLM Land 21

Table 10. BLM RE Projects Potential Contribution to 2020 RPS Generation Requirements 22

Table 11. BLM Proposed Geothermal Projects beyond 20 Miles of Transmission 33

Table 12. BLM Proposed Solar Projects beyond 20 Miles of Transmission 34

Table 13. BLM Proposed Wind Projects beyond 20 Miles of Transmission 35

Background

Over the last 10 years, the U.S. Department of the Interior (DOI) and the Bureau of Land Management (BLM) have been working with local communities, state regulators, industry, and other federal agencies to identify and provide suitable sites for environmentally sound development of renewable energy on public lands. Renewable energy projects on BLM-managed lands include wind, solar, geothermal, and biomass projects. Also included are projects to site for transmission facilities that deliver this power to the consumer.

Although most federal land areas have some renewable energy potential, the majority of resource potential on federal lands is concentrated in western states where the bulk of federal lands are located. Of the 247.9 million acres of public land managed by the BLM, 174.2 million acres (70%) are located in the 11 western states considered in this study (DOI/BLM 2010a). As of December 2010, the BLM has identified 22 million acres of public lands in six states that have high solar energy potential and 20.6 million acres of public lands in 11 states with wind potential. In addition, the BLM has delegated authority for leasing 149 million acres of public lands (plus an additional 100 million acres of National Forest lands) with geothermal potential (DOI/BLM 2010b).[6]

The siting of renewable energy generation facilities is largely dictated by the location of the highest quality resources. Consequently, a significant increase in the use of renewable energy is dependent on expansion of the transmission grid. Federal lands are expected to have a major role in the transmission of renewable electricity from remote wind, solar and geothermal projects to the nation's population centers.

Current Renewable Energy Markets in the Western United States

This section summarizes current renewable energy markets in the western region – both supply (generation or capacity) and demand (sales in GWh) – for total markets, RPS-driven compliance markets, and voluntary consumer-driven markets.

Total Markets

In 2010, the western region accounted for 731,000 GWh (EIA 2011d), or 18% of total U.S. electricity generation. Electricity consumption and generation for the study area by state is shown in Figure 1. Electricity consumption in California accounts for 38% of the total electricity demand in the western states, but California generates only 28% of the electricity produced in the western states.

---

[6] These are lands that have solar, wind or geothermal resources that are technically suited for commercial development. This does not imply that these lands are economically or environmentally suitable for development.

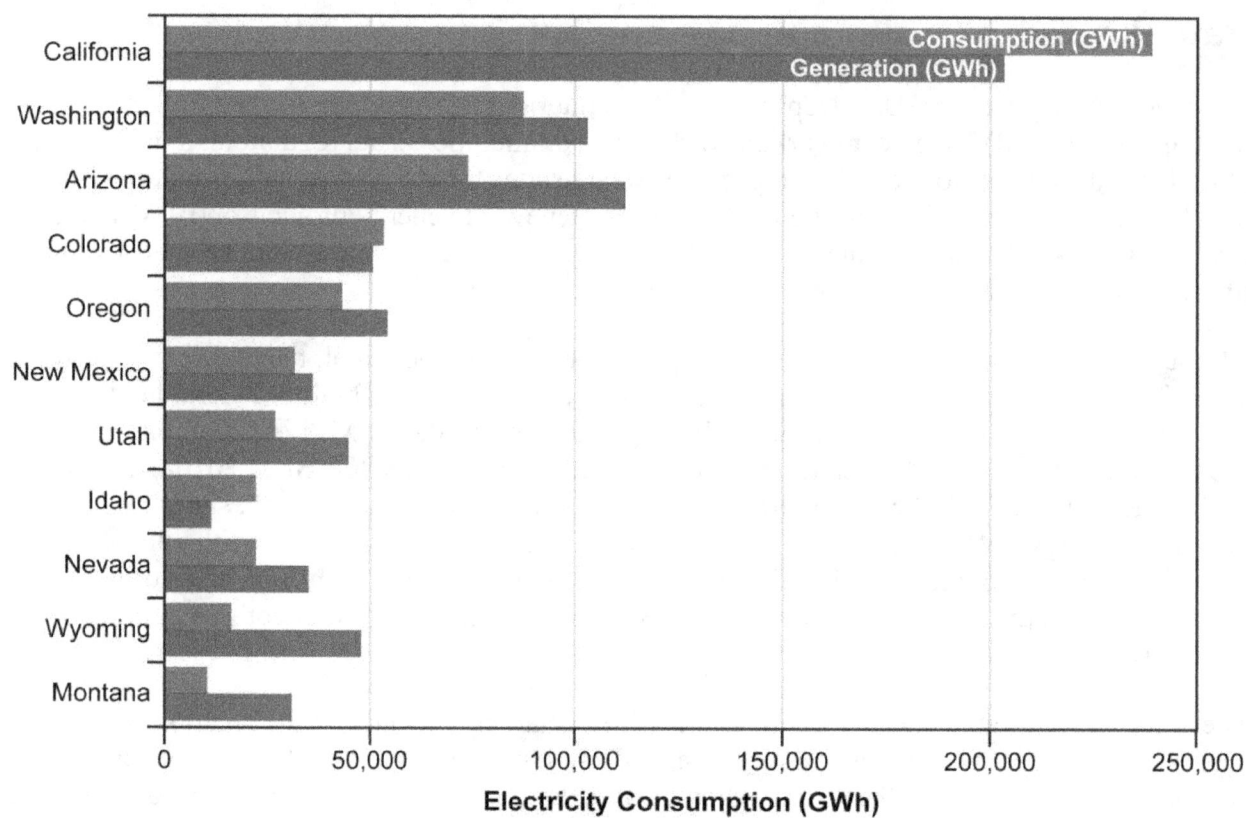

Figure 1: Total (all fuels) state electricity consumption and generation in 2010.

Source: EIA 2011c and EIA 2011d

The overall generation mix for states in the WECC is shown in Figure 2. Coal is the largest single resource, and together with natural gas accounts for 60% of the region's generation. Hydropower is 22% of the generation; nuclear, 10%; and RE, 7%.

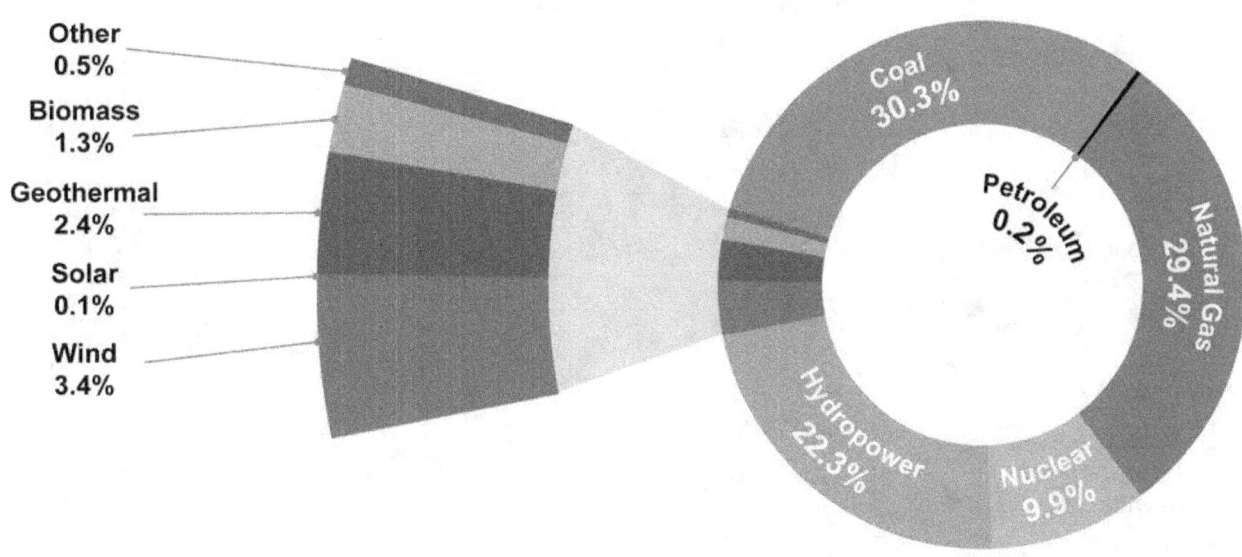

Figure 2. WECC generation mix in 2010

Source: EIA 2011d

The estimated 2010 electricity generation by fuel for each western state is shown in Figure 3. The main generating source in eight of the 11 states is either coal or a combination of coal and natural gas.

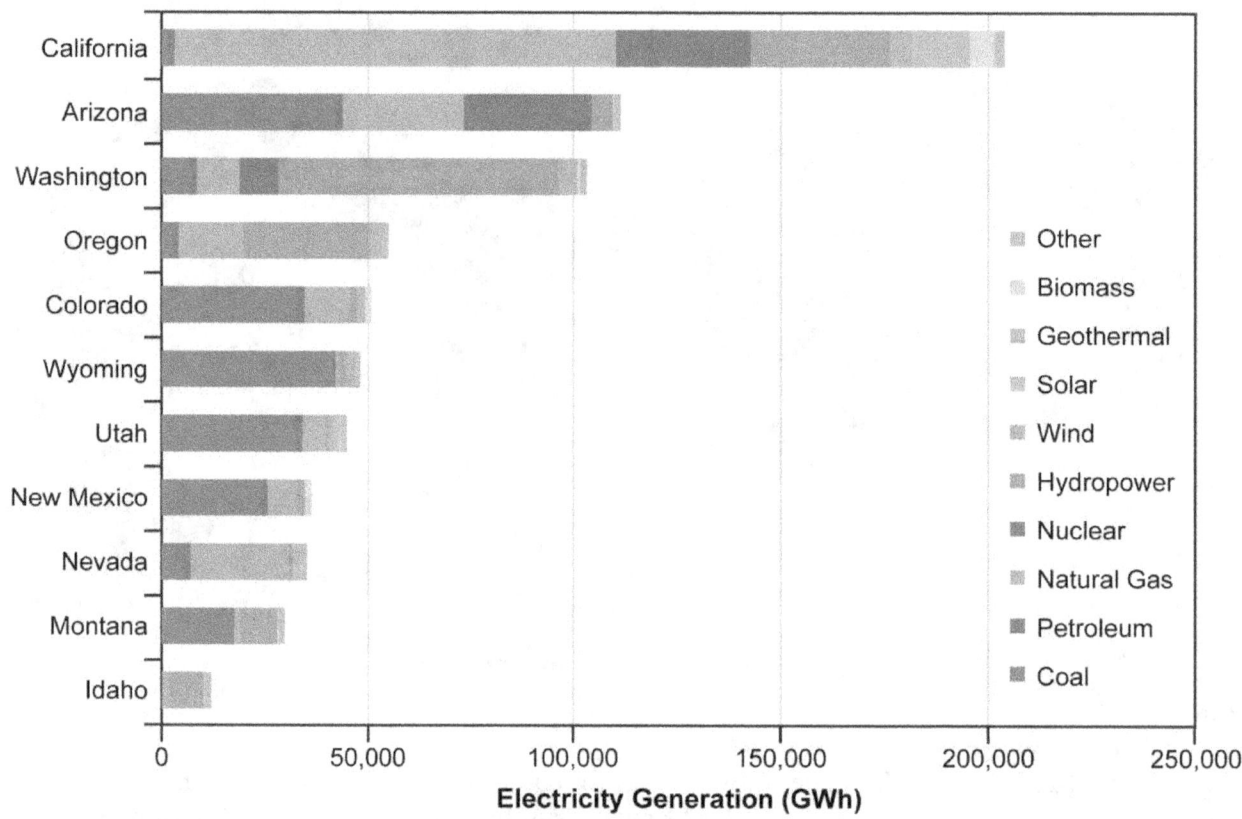

Figure 3. Estimated state electricity generation by fuel source for 2010

Source: EIA 2011d

In 2010, the 11 western states had a total estimated non-hydropower-RE nameplate capacity of 18,429 MW (Table 1). The majority of this capacity is from wind (64%). Geothermal is 19%, biomass is 12%, and solar (both PV and CSP) is 5%. At 7,880 MW, California has the highest non-hydropower renewable energy nameplate capacity in the western region, more than three times the next closest states of Washington and Oregon, who each have about 2,500 MW. (The number for PV does not include distributed generation projects.)

Table 1. State Renewable Nameplate Capacity (MW) as of December 2011

	Arizona	California	Colorado	Idaho	Montana	Nevada	New Mexico	Oregon	Utah	Washington	Wyoming	Total WECC	Percent
Wind (MW)	128	3,124	1,297	288	385	0	701	2,204	222	2,086	1,418	**11,854**	**64%**
Biomass (MW)	40	1,200	21	136	19	0	7	346	12	461	0	**2,241**	**12%**
Solar – PV (MW)	23	175	54	0	0	77	30	8	0	1	0	**368**	**2%**
Solar – CSP (MW)	3	413	0	0	0	76	0	0	0	0	0	**491**	**3%**
Geothermal (MW)	0	2,968	0	18	0	437	0	0	52	0	0	**3,475**	**19%**
Total Renewable Energy (Excluding Hydropower) (MW)	**194**	**7,880**	**1,372**	**441**	**404**	**589**	**738**	**2,558**	**287**	**2,547**	**1,419**	**18,429**	**100%**

Source: SNL 2011c

Renewable energy markets are driven by multiple factors, including

- RPS requirements passed within each state

- Voluntary consumer demand markets for RE or renewable energy certificates (RECs)

- Other less well-defined factors or markets that encourage utilities to purchase or produce renewable energy. This could include the procurement of RE by utilities on a strictly cost-competitive basis through the more traditional procurement of power through the integrated resource planning process; federal directives regarding requirements for renewable generation on federal sites of all types; state and federal tax credit and similar incentives; changes in RPS requirements in the future; and desire of states to diversify their economies and develop green energy jobs.

RPS-Driven (Compliance) Markets

This analysis is based primarily on RPS-driven requirements because these data are most readily available. The demand for RE in the western states required to meet mandatory state RPS requirements for 2010, based on utility compliance filings, is presented in Table 2 (Barbose 2011). Note that hydropower is excluded from this table. Some types of hydropower do meet RSP requirements in certain states, such as small hydro and low-impact hydro, but data available are insufficient to include only RPS-qualifying hydropower, so hydropower was excluded for simplicity. Lastly, Table 2 indicates that the WECC states in total have sufficient generation today to meet the overall RPS requirements, but drawing that conclusion about individual states from this data would be less appropriate. A state can often meet all or part of their RPS requirements with out-of-state supply, such that more detailed analysis of each state's compliance status is necessary.

Table 2. Renewable Energy Generation Required for Mandatory State RPS in 2010 Compared to Actual Renewable Energy Generation in 2010 Excluding Hydropower

		Arizona	California	Colorado	Idaho	Montana	Nevada	New Mexico	Oregon	Utah	Washington	Wyoming	Total WECC
2010 Actual RE Generation (GWh) (Plants Located within the State)	2010 RPS Requirement (GWh)	1,016	41,902	1,700	(a)	692	3,381	852	(a)	(a)	(a)	(a)	49,544
	Total (Excluding Hydropower[1])	319	25,450	3,555	1,014	1,027	2,287	1,855	4,757	3,369	6,617	3,247	53,496
	Biomass	168	6,002	60	501	97	0	14	837	56	1,872	0	9,608
	Geothermal	0	12,600	0	72	0	2,070	0	0	2,865	0	0	17,607
	Solar	16	769	42	0	0	217	9	0	0	0	0	1,053
	Wind	135	6,079	3,452	441	930	0	1,832	3,920	448	4,745	3,247	25,228
	Hydro	6,622	33,431	1,578	9,154	9,415	2,157	217	30,542	696	68,288	1,024	163,125

Sources: Barbose 2010, Barbose 2011, EIA 2011a

Notes:
1. Oregon's RPS requirements begin in 2011; Washington's begin in 2012; Idaho, Utah, and Wyoming do not have RPSs.
2. Some types of hydropower do meet RSP requirements in certain states, such as small hydro and low-impact hydro, but data was insufficient to select only RPS-qualifying hydropower, so hydropower was excluded.
3. Actual generation is categorized by state according to the state where the power plant is located. However, RPS requirements are often met from out-of-state purchases.

Voluntary (Consumer-Driven) Markets

Voluntary purchases of renewable energy are generally in addition to renewable energy used to meet RPS targets, sometimes called compliance markets (Heeter 2011). Voluntary purchases represent those made by consumers who are willing to pay a premium for green power to support renewable energy that would not have been otherwise supported through RPS regulatory requirements or who might want to go beyond a state's RPS.

Figure 4 shows findings of a recent study that documents a steady growth in voluntary consumer demand markets over the last five years (Heeter 2011), to a total in 2010 of 35,600 GWh. The only ready source for state-by-state data is from the Energy Information Administration (EIA); EIA reported that in 2010, the portions of the voluntary markets that they track totaled 3,200 GWh in the western states (see Figure 5). A comparison to the estimated 2010 RPS-driven requirement of 49,500 GWh (Table 2) in WECC states suggests that while voluntary markets are not as large as RPS-driven markets, they could be substantial markets to consider in the western region. However, no projections of possible voluntary markets in 2020 have been published, so voluntary markets were not quantitatively considered in the supply and demand calculations for 2020. Also, voluntary demand can be met by purchasing renewable energy credits from any state, including states outside WECC. Therefore, including voluntary market projections as an element of a supply/demand gap calculation for WECC could be misleading.

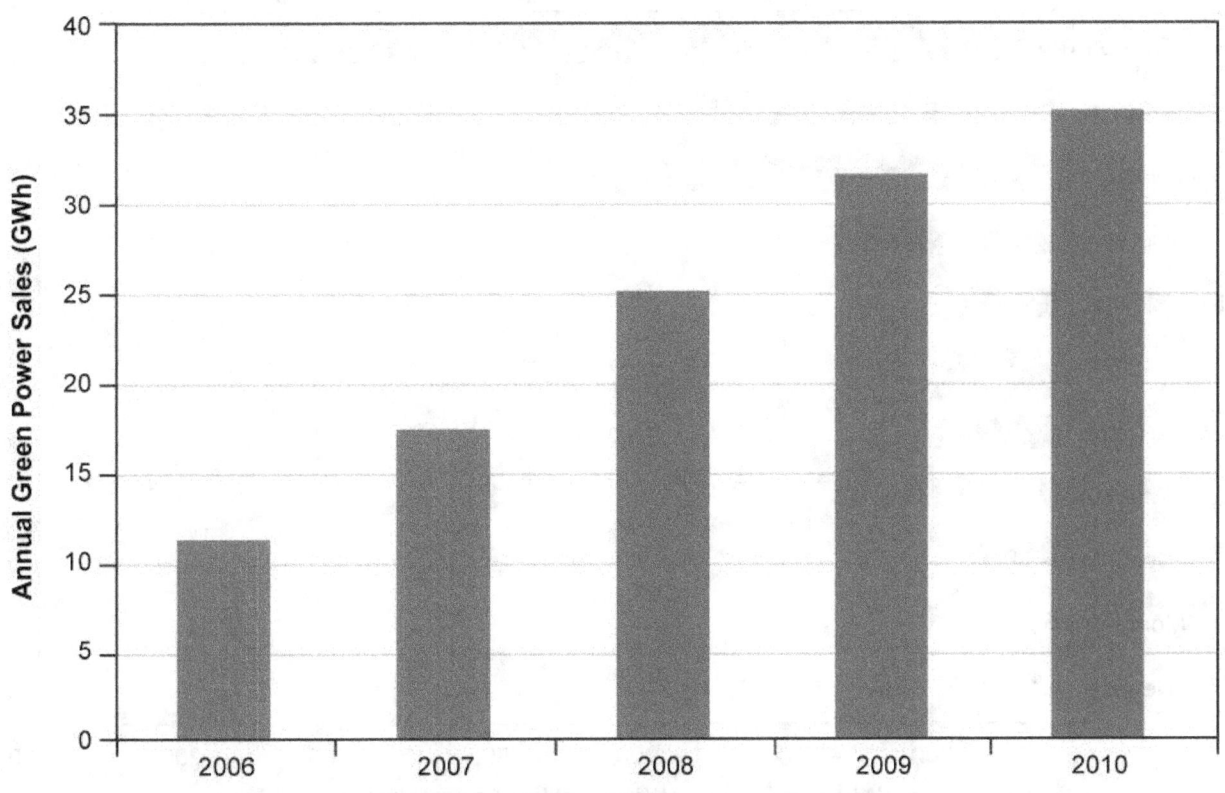

Figure 4. Growth in national voluntary green power market segments, 2006-2010.

Source: Heeter 2011

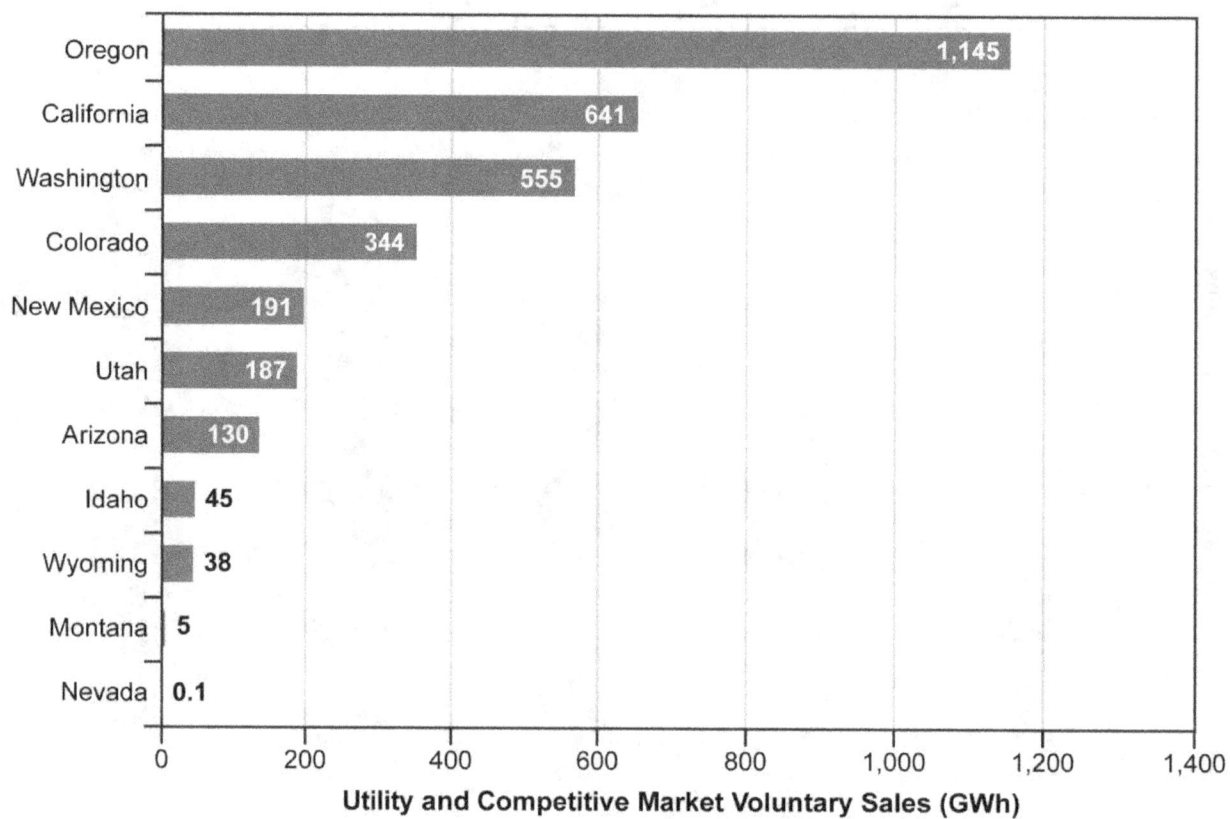

Figure 5. Voluntary utility green pricing and competitive market power sales by state in 2010.

Source: EIA 2011e

Projections for Estimating the Supply-Demand Gap in the Western Region

One of the primary purposes of this analysis is to estimate the gap between the RE supply and demand in the western states. Because of the better availability of data, this analysis gives a special emphasis on the gap in capacity needed to meet current RPS requirements. This gap estimate requires three sets of data: demand in terms of the capacity required by the RPS policies of each state, supply in terms of the current (2010) RE capacity that qualifies for each state's RPS (Table 4), and supply in terms of the capacity of planned projects that qualify to meet each state's RPS. The best available data for each of these three data sets is discussed in this section.

Demand Projections, 2010-2035

The projected total statewide electricity demand (in GWh) for individual western states through 2035 is presented in Figure 6 (Barbose 2011). Actual retail sales data are sourced from EIA through 2008 from the Retail Sales of Electricity by State by Sector by Provider, 1990-2010 spreadsheet (http://www.eia.gov/cneaf/electricity/epa/sales_state.xls). EIA collects these data annually through Form EIA-861, the Annual Electric Power Industry Report. This form compiles operational data, revenues, sales and customer counts among other data for each utility. Various growth rates are applied by LBNL to forecast sales by state through 2035.

8

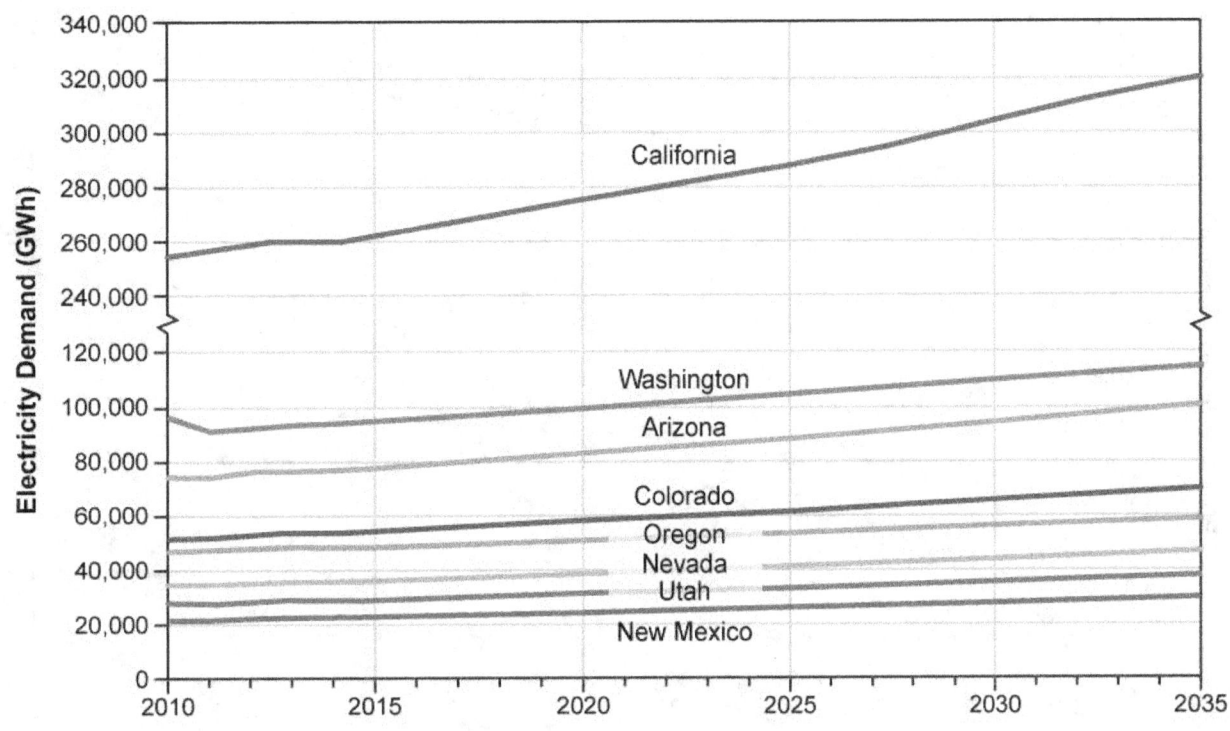

Figure 6. Projected electricity demand by state.

Source: Barbose 2011

Not all retail sales may necessarily be counted toward a state's RPS requirements. For example, the sales from investor-owned utilities may count toward a state's RPS while those from a municipal utility or a cooperative may not. RPS requirements vary markedly state to state. The RPS applicable load is calculated for each state for each category of supplier (investor owned utilities, municipal utilities, cooperatives, power marketers, etc.). RPS target percentages (based on a target schedule) are applied to the RPS applicable load to reach a total RPS target in GWh. From there, solar set-aside and distributed generation set-aside target percentages are applied to estimate the fraction of total GWh that will be solar and distributed generation. In summary, given the details of each state's RPS, it is possible to estimate the amount of this demand to which each state's RPS could be applied, as shown in Figure- 7 (Barbose 2011). The amount of a state's total demand (Figure 6) that can be applied to the state's RPS (Figure 7) varies from state to state.

9

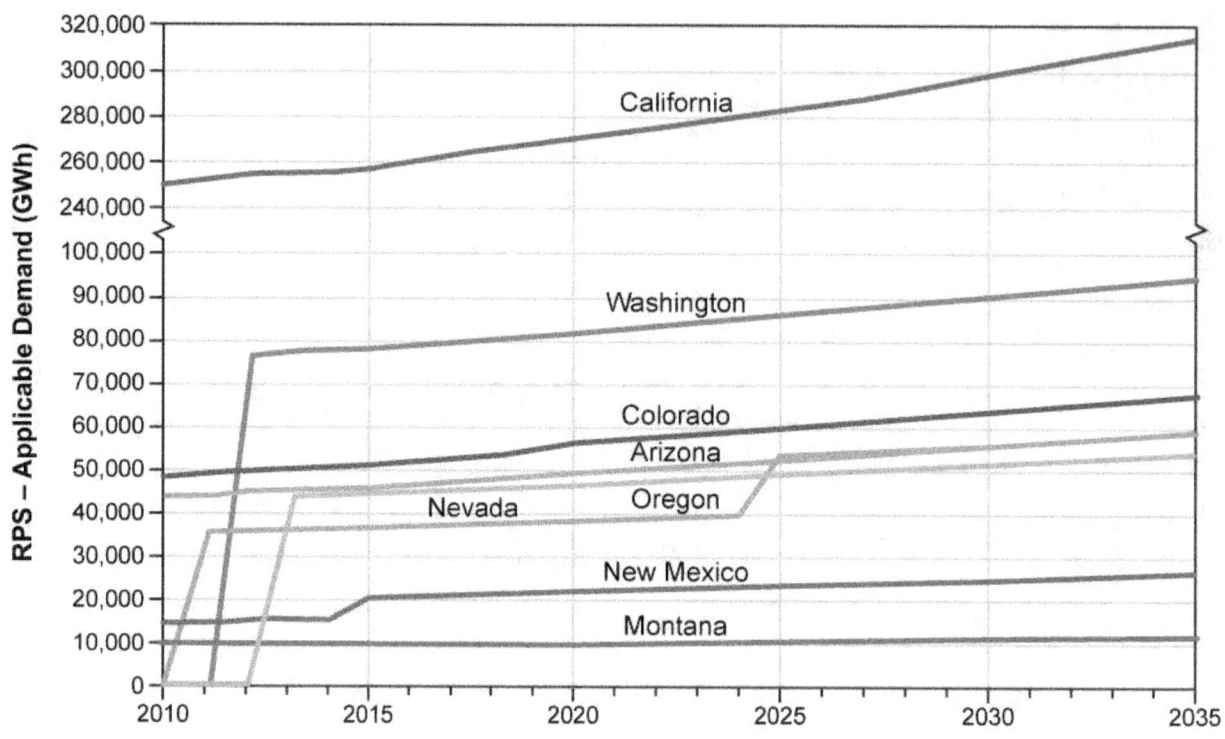

Figure 7. Projected RPS-applicable demand by state – WECC states.

Source: Barbose 2011

Once the appropriate amount of demand is determined, the renewable energy generation specifically required to meet each state's RPS can be estimated, as in Figure 8 (Barbose 2011). The curves are usually stepped according to when particular levels of RE participation are required in future years for each state's unique RPS.

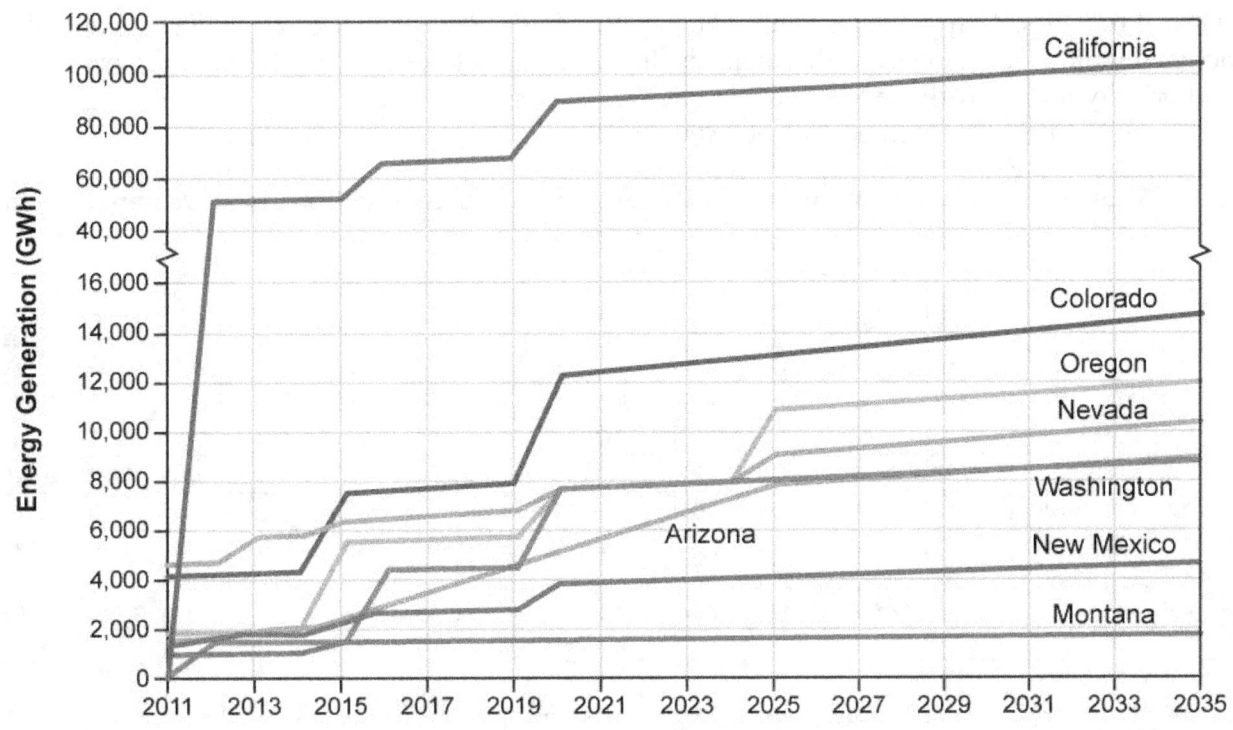

Figure 8. Projected total renewable energy generation required to meet RPS – WECC states.

Source: Barbose 2011

Supply: Planned Projects by 2020

In the 11 WECC states, SNL databases as of November 2011 (SNL 2011) indicate that approximately 22,700 MW of new renewable generation capacity is currently planned, with 18,800 MW of that capacity coming from non-hydropower sources. These figures (Table 3) are conservative in that they only include planned RE projects categorized as "Advanced Development" or "Under Construction." ("Advanced Development," using the SNL Financial definition, includes projects that meet two of the following criteria: have a signed power purchase agreement, obtained financing, procured turbines or other major parts, hired an engineering, procurement and construction (EPC) contractor, or obtained all permits.) It should be noted that even if a project has a power purchase agreement, has obtained permit approvals, or has closed on financing, there is no guarantee that the project will be built.

The output from planned projects was assigned to states based on the location of the power purchaser, if known; if not known, the project was assigned to the state where it is planned to be located. About 68% of the planned capacity of the WECC states overall has a known power purchaser. Most (about 98%) of the planned capacity in California and Nevada is known; about 68% is known for Arizona, Colorado, and Idaho; and the others have zero to a few percents of their planned capacity with known power purchasers. Some larger power purchasers sell power to states outside where they are located, so a further refinement was made in assigning states to planned capacity for those individual purchasers. In those cases, a recent analysis (Hurlbut 2011) examining each utility's power purchase agreements was used to apportion the output more accurately. Projects with unknown power purchasers were counted in the state in which the project will be physically located.

11

Many of the projects approved or authorized by BLM as of this date (see Appendix) are already included in this total. However, this analysis did not cross-reference every BLM approved or high priority project with the SNL data. Also, this analysis did not analyze how many of the approved BLM projects will help meet a state's RPS.

Table 3. Estimated Capacity of Currently Planned Renewable Energy Projects in Western Region

	Arizona	California	Colorado	Idaho	Montana	Nevada	New Mexico	Oregon	Utah	Washington	Wyoming	Total
Wind (MW)	229	2,842	29	472	300	152	125	655	0	1,160	2,841	8,805
Biomass (MW)	0	98	0	9	0	14	0	20	0	0	0	141
Solar (MW)	703	6,659	62	20	0	1,053	5	14	0	75	0	8,590
Geothermal (MW)	113	769	0	63	0	155	0	174	0	10	0	1,283
Hydro (MW)	0	1,756	0	0	1	0	0	2,013	0	58	5	3,833
Renewable Energy Total Including Hydropower (MW)	1,045	12,124	91	564	301	1,374	130	2,876	0	1,303	2,846	22,653
Renewable Energy Total Excluding Hydropower (MW)	1,045	10,367	91	564	300	1,374	130	864	0	1,245	2,841	18,820

Source: SNL 2011c

Demand: RPS Requirements in 2020

The remainder of the analysis will focus on 2020 as a near-term point in time for estimating the gap between supply and demand. Figure 8 gives the generation required to meet RPS requirements in 2020. To convert that generation into RE plant nameplate capacities, a choice must be made about which capacity factors[7] to use. Though these vary from state to state, PV capacity factors tend to be around 26% and wind capacity factors are around 35%. For states that specify a solar or wind fuel-source or distributed-generation set-aside, appropriate capacity factors were used (Barbose 2011). In three states (California, Colorado and Nevada), expectations of the types of projects were known such that specific capacity factors could be used for all new capacity, not just set-asides. In the remaining five states with an RPS, most RE generation is not specified by fuel source in the RPS. The choice of capacity factor in those situations requires predicting what technologies might be built, and suggests three different approaches. Generally, LBNL used 50% as a generic capacity factor in these cases. This analysis quotes the LBNL estimates for 2020 (labeled "B" in Tables 5 and 6). In addition, this analysis calculates two alternatives: one with a lower capacity factor, which will result in a higher need for RE capacity (labeled "C" in Tables 5 and 6), and one with a higher capacity factor, which will result in a lower need for RE capacity (labeled "A" in Tables 5 and 6).

---

[7] The capacity factor is the fraction or percentage of time that a renewable energy project generates electricity. The higher the capacity factor, the more time the project operates, and the lower the size or capacity of the plant needed to generate that amount of electricity. Capacity factors vary with the fuel source, generation technology, and available resource or location.

Table 4 shows how these high- and low-capacity factors, resulting in low and high-capacity requirements respectively, were estimated for unspecified fuel sources. Capacity factors were used from actual RE projects in WECC states taken from the SNL database (November 2011) for the period 2008-2010, excluding projects less than four years in operation and those without valid capacity factors. To get weighted average capacity factors for each state, the average capacity factor for each technology was calculated from this set of data and then applied against the actual installed capacity in 2010 (Table 1). Among the WECC states, Colorado projects have the lowest weighted average capacity factor (26%), and Nevada the highest (53%). These lowest and highest actual weighted average values were used as the boundary conditions for further calculations.[8]

Table 4. Calculations to Arrive at Lowest (26%) and Highest (53%) Capacity Factors for Use in 2020 Estimates (Tables 5 and 6)

Weighted Average Capacity Factor (%) Using Capacity Factors and 2010 Installed Capacities from WECC States					
Wind	26.6%	Solar Photovoltaics	17.7%	Geothermal	66.4%
Biomass	62.5%	Concentrated Solar Power	21.3%		

	Arizona	California	Colorado	Idaho	Montana	Nevada	New Mexico	Oregon	Utah	Washington	Wyoming
Weighted Average Capacity Factor	27.6%	42.7%	26.3%	37.3%	28.1%	53.4%	26.4%	32.0%	33.8%	32.7%	26.6%

Table 5 lists the RE generation requirements in 2020 taken from Figure 8 for meeting RPS requirements, along with three estimates for the RE capacity required to produce this electricity. The high (53%) and low (26%) capacity factors from Table 5 were used to calculate Estimates A and C for all future RE requirements in the LBNL dataset where fuel source was not specified. Estimate B used LBNL's approach to capacity factors for all non-specified RE requirements (Barbose 2011). For simplicity, the capacity factors of the relatively small generation requirements for set-asides and distributed generation were taken from LBNL (LBNL 2011) without alternative approaches. Table 5 shows that the total capacity needed in WECC states for RPS requirements in 2020 could range from 28,500 to 45,500 MW, depending on these assumptions about capacity factors.

---

[8] Concentrated solar thermal power plants (CSP) with storage may have capacity factors as high as 70%, and may be part of the future WECC installations. While this figure was not explicitly included in the calculation of the upper boundary for the capacity factor, a modest amount of CSP with storage could logically be accommodated within the upper boundary for the weighted average capacity factor (which corresponds to the lowest required capacity, and therefore the potential oversupply situation in 2020.)

Table 5. Estimation of Capacity (MW) Based on Generation (GWh) Needed to Meet RPS Targets in 2020

	Capacity Factor for Non-specified RE Generation[1]	Arizona	California	Colorado	Idaho	Montana	Nevada	New Mexico	Oregon	Utah	Washington	Wyoming	Total WECC
		10%	33%	30% IOU Total, 3% IOU DG,10% POU	No RPS	15%	22%	20% IOU, 10% Co-Op	20% large utilities, 5% small utilities	No RPS	15%	No RPS	
Total RPS Qualifying Generation Needed in 2020 (GWh) (Barbose 2011)	n/a	4,882	89,259	12,227	n/a	1,508	7,464	3,814	7,610	n/a	7,607	n/a	134,372
Utility Scale and Non Set Aside Capacity Required in 2020 to Meet RPS — A High Capacity Factor, Low Capacity (MW)	53%	735	18,653	2,741	n/a	325	783	446	1,713	n/a	1,807	n/a	27,205
B LBNL Capacity Factors (Barbose 2011) (MW)	Various	771	27,944	2,741	n/a	382	783	473	1,791	n/a	n/a	n/a	34,886
C Low Capacity Factor, High Capacity (MW)	26%	1,499	31,354	2,741	n/a	662	783	909	3,059	n/a	3,238	n/a	44,247
Distributed Generation and Set Aside Capacity in 2020 (MW) (Barbase 2011)	n/a	644	n/a	n/a	n/a	n/a	112	512	20	n/a	n/a	n/a	1,288
Total Capacity Required in 2020 to Meet RPS — A High Capacity Factor, Low Capacity (MW)	53%	1,380	18,653	2,741	n/a	325	896	958	1,733	n/a	1,807	n/a	28,493
B LBNL Capacity Factors (Barbose 2011) (MW)	Various	1,415	27,944	2,741	n/a	382	896	985	1,811	n/a	2,952	n/a	39,126
C Low Capacity Factor, High Capacity (MW)	26%	2,144	31,354	2,741	n/a	662	896	1,421	3,079	n/a	3,238	n/a	45,535

Projected Gap in RE Capacity in 2020

For estimating the potential gap between RE supply and RE demand in the western states, these various market projections were compared:

- RE supply was considered in terms of what RE capacity is installed through approximately December 2011 in the western states (18,400 MW).

- RE supply was also considered in terms of what RE capacity is planned to be added as of November 2011, which will come online in the period between December 2011 and 2020 (18,800 MW). Table 6 row 2 is repeated from Table 3. This data represents all planned capacity, whether driven by RPS or by any other market factor. The sum of actual installed capacity today and planned capacity is about 37,000 MW, as shown in Table 6.

- RE demand in Table 6 rows 4-6 is taken from Table 5 above, and represents only RPS-driven demand. It is worth noting that RE demand will also be driven by voluntary markets, other federal and state incentives, changes in transmission availability, economic considerations at individual utilities as reflected in their normal integrated resource planning processes, and other factors. These and other market factors could increase or decrease these market estimates.

- The gap between RE supply and demand in the western states, shown in Table 6, reflects a minimum gap. With the data available for this analysis at this time, projections of some key market drivers are not available, as noted immediately above.

Table 6. Projected Oversupply or Unmet Demand for RE Capacity in 2020 in Western States

			Arizona	California	Colorado	Idaho	Montana	Nevada	New Mexico	Oregon	Utah	Washington	Wyoming	Total WECC
SUPPLY	As of December 2011 – Not Including Hydropower[4]		Existing Renewable Energy Plant Capacity (MW)[1] (SNL 2011c)	194	7,880	1,372	441	404	589	738	2,558	287	2,547	1,419	18,429
			Planned Renewable Energy Projects Capacity (MW) (SNL 2011a)	1,045	10,367	91	564	300	1,374	130	864	0	1,245	2,841	18,820
			Sum of Existing and Planned Renewable Energy Capacity (MW)	1,239	18,247	1,462	1,005	704	1,963	868	3,422	287	3,792	4,260	37,249
DEMAND	Capacity Required in 2020 to Meet Current RPS Requirements (MW)[2]	A	High Capacity Factor, Low Capacity (MW)	1,380	18,653	2,741	0	325	958	896	1,733	0	1,807	0	28,493
		B	LBNL Capacity Factors (Barbose 2011) (MW)	1,415	27,944	2,741	0	382	896	985	1,811	0	2,952	0	39,126
		C	Low Capacity Factor, High Capacity (MW)	2,144	31,354	2,741	0	662	1,421	896	3,079	0	3,238	0	45,535
GAP	Oversupply or (Unmet Demand) in 2020[3]	A	High Capacity Factor, Low Capacity (MW)	(141)	(406)	(1,279)	1.005	379	1,005	(28)	1,689	287	1,985	4,260	8,756
		B	LBNL Capacity Factors (Barbose 2011) (MW)	(176)	(9,697)	(1,279)	1,005	322	1,067	(117)	1,611	287	840	4,260	(1,877)
		C	Low Capacity Factor, High Capacity (MW)	(905)	(13,107)	(1,279)	1,005	42	542	(28)	343	287	554	4,260	(8,286)

Notes:
1. These data are from Table 1 (SNL 2011c), and do not distinguish between projects responding to RPS and projects responding to other market drivers.
2. See text for description of three capacity factor scenarios.
3. Various market factors could increase or decrease these market estimates.
4. Some of the 3,833 MW of planned hydropower will likely qualify for an RPS; however, RPS regulations on hydropower are complicated, and for simplicity, hydropower was omitted from this estimate.

Another recent study (Bird 2010) also examines the supply-demand balance in RE markets by region, and concludes that the western region has ample RE supply in 2015 compared to RPS and some voluntary market demand. However, this conclusion was based on actual data from 2006-2008, and used market forecasts and regression analysis for supply projections. In contrast, the current analysis for supply projections uses actual installed RE projects through 2010 and projections based on SNL's database.

BLM Renewable Energy Activities

Under Title II, Section 211 of the Energy Policy Act of 2005 (EPAct 2005), the Secretary of the Interior is required to approve 10,000 MW of non-hydropower renewable energy to be located on public lands by 2015. On January 16, 2009, former Interior Secretary Dirk Kempthorne issued Secretarial Order 3283 that authorized the BLM to establish renewable energy coordination offices (RECOs) to expedite the permitting of utility-scale renewable energy projects on public lands (DOI/BLM 2009). At this time, BLM had authorized approximately 1,800 MW of renewable projects (0 MW solar, 576 MW wind, 1,300 MW geothermal) and had a backlog of some 400 pending renewable energy applications.

In March 2009, Secretary Salazar issued Secretarial Order 3285 (SO 3285), which established the production, development and delivery of renewable energy as one of the Department of the Interior's highest priorities. SO 3285 also required DOI bureaus and agencies to work cooperatively with each other as well as with other federal, state and local agencies to encourage the timely and responsible development of renewable energy and transmission while protecting natural resources. In addition, SO 3285 established the Task Force on Energy and Climate Change within the DOI, which, amongst other responsibilities, was charged with establishing Department-wide processes, practices and strategies to implement the environmentally responsible development of renewable energy and transmission on public lands (DOI 2009).

Over the remainder of 2009, DOI and the BLM announced the opening and staffing of Renewable Energy Coordination Offices in Arizona, Nevada, California and Wyoming. The BLM also established and staffed renewable energy permitting teams in Colorado, Idaho, Montana, New Mexico, Oregon and Utah.

In June 2009, in order for developers to take advantage of certain American Recovery and Reinvestment Act incentives (primarily the Department of Energy's Loan Guarantee Program and the Treasury Department's 1603 Program), the BLM announced that 31 wind, solar and geothermal projects that met certain criteria would be placed on a "Fast-Track" list for full environmental analysis and public review by the end of 2010. BLM continued this process in 2011 and 2012, with the identification and focus on several dozen more "high priority projects."

Since June 2009, the BLM has approved 27 renewable energy projects, totaling approximately 6,600 MW of new, utility-scale RE to be constructed on BLM lands[9]. See Appendix A for a listing of all the BLM high priority projects approved during the past two years, and Appendix B for the 2012 list of high priority projects.

Programmatic Environmental Impact Statements

Over the last six years, the BLM has systematically identified potential locations for geothermal, wind and solar resources on federal lands based on a variety of constraints—environmental, legal, policy—through the Programmatic Environmental Impact Statement (PEIS) process. These PEISs evaluated utility-scale RE development, developed programs and guidance for environmental policies and mitigation strategies for RE projects, and amended relevant BLM land use plans. The following text summarizes these efforts:

- The draft Solar Energy Development PEIS (DOI/BLM 2010) identified 24 proposed solar energy zones in six western states as priority development areas for utility-scale solar energy facilities. A supplement to the draft PEIS, released October 2011, reduced that number to 17 solar energy zones with additional guidance on proposing new solar energy zones and developing outside of solar energy zones. A solar energy zone is defined by the BLM as an area with few impediments to utility-scale production of solar energy where BLM would prioritize solar energy and associated transmission infrastructure development. The final Solar PEIS is expected to be published in late 2012.

---

[9] As of December 30, 2011.

- The Wind Energy Development PEIS (DOI/BLM 2005) evaluated the potential impacts associated with wind energy development on federal lands, including the adoption of policies and best management practices and the amendment of 52 BLM land-use plans to address wind energy development.

- The Geothermal Resources Leasing PEIS (DOI/BLM 2008) focused on analyzing and expediting the leasing of BLM- and Forest Service-administered lands with high potential for renewable geothermal resources in 11 western states and Alaska.

The BLM's Reasonably Foreseeable Development Scenario – Solar PEIS

In the December 2010 draft Solar PEIS, the BLM included a Reasonably Foreseeable Development Scenario (RFDS) in order to help define the possible extent of solar development within the PEIS' six-state planning area. The RFDS is an estimate of the number of megawatts of solar energy that could be developed on BLM lands in the PEIS' six southwestern states by 2030, as well as a corresponding estimate of the number of acres necessary to produce that amount of power.

The RFDS ensures the usefulness of the PEIS through 2030. If demand for BLM-managed land for solar projects exceeded the scope of the PEIS' Record of Decision, the BLM would be limited in its ability to meet that additional demand without doing additional planning. Therefore, the BLM's projections of the demand for utility-scale solar in the six southwestern states are based on the following "upper bound" assumptions:

- 50% of the total RPS for each state would come from solar energy

- 75% of solar energy development in the six states would occur on BLM-managed land

- Each megawatt of solar energy would require nine acres of land to develop (five acres for parabolic trough CSP).

Table 7 shows the BLM's RFDS estimates for solar energy by 2030 for each state in addition to the overall range of numbers for projected RPS demand in 2020 (from Table 6). While the BLM values are estimated through 2030 and this paper only looks at demand through 2020, it should be noted that the BLM RFDS of nearly 24,000 MW of solar development on BLM lands is nearly equivalent to the minimum total 2020 RPS demand for the six states, and is 60% of the projected maximum RPS demand. While not an apples-to-apples comparison due to the timeframe differences, further analysis such as described below shows that it may not be realistic to assume that so much of the RPS demand will be filled by solar energy development, or by solar development on public lands.

Table 7. Six States: Comparison of Solar PEIS Projections for 2030 and Projections Based on RPS-Driven Demand in 2020

		Arizona	California	Colorado	Nevada	New Mexico	Utah	Total
Solar PEIS Reasonably Foreseeable Development Scenario by 2030	BLM Solar (MW)	2,424	15,421	2,194	1,701	833	1,219	**23,792**
	Total Solar (MW)	3,232	20,561	2,925	2,268	1,111	1,625	**31,722**
Projected 2020 Total RPS-Demand Across All Renewable Technologies (Table 5)	Minimum MW (High Capacity Factor, Low Capacity)	1,380	18,653	2,741	896	958	0	**24,628**
	Maximum MW (Low Capacity Factor, High Capacity)	2,144	31,354	2,741	896	1,421	0	**38,556**

In terms of available land, the BLM recently released a supplement to the draft Solar PEIS. Though it did not alter the values in the RFDS, the BLM acknowledged that under at least one alternative presented in the PEIS (the Solar-Energy-Zone-only alternative), there would not be enough land available in some states to handle the amount of development projected in the RFDS. However, the PEIS identifies methods for adding new solar energy zones, and recognizes other planning efforts related to solar development on public lands that could meet this gap, including Arizona's Restoration Design Energy Project for building solar and wind projects on previously disturbed lands and California's DRECP process.

Supplement to the Draft Solar PEIS – Transmission Analysis
In the October 2011 supplement to the draft Solar PEIS, the BLM noted that many commenters – agency, industry, and environmental groups – expressed concern about the lack of analysis of transmission, especially as it relates to connecting the solar energy zones to existing transmission infrastructure. While the supplement does not include a robust transmission analysis, it does detail a specific course of analysis that the BLM will undertake in the final Solar PEIS. That analysis will provide more information to agencies and interested groups regarding transmission access and cost issues with the approved solar energy zones. It will also detail what new transmission might be needed to support development in the solar energy zones. In preparation for that analysis, the supplement details the limitations regarding an accurate prediction for transmission needs for solar projects, methods the BLM will employ when conducting the transmission analysis, and a sample analysis for the Brenda solar energy zone located in Arizona.

BLM's Priority Projects
In June 2009, the BLM established and implemented a process to prioritize proposed RE projects on federal lands. The process requires early coordination and careful review of proposed renewable energy projects with federal, state, tribal and local government agencies before committing significant resources to the processing of solar and wind energy development right of way applications. This requirement assists the BLM in identifying and prioritizing those applications that have the fewest resource conflicts and the greatest likelihood of success in the permitting process. In addition, to be a priority project, a company must demonstrate to the BLM that the project has progressed far enough to formally start the environmental review and public participation process, as well as have the potential to be cleared for approval by the end of the

year in which the project is given priority status. The projects must also be sited in an area that minimizes impacts to the environment.

In 2010, the BLM reviewed a number of renewable projects, approving nine solar projects (3,600 MW total), one wind project (150 MW) and two geothermal projects (nearly 80 MW). For 2011, the BLM identified 18 priority RE projects including nine solar projects (including those connected action projects located on private and tribal lands) (2,673 MW), four wind projects, including two connected action projects, (854 MW), and five geothermal projects (312 MW) throughout the western region. Many of the 2011 projects have already been approved and are included in the list in Appendix A. Several others are likely to be approved in January 2012. The final list of 2012 priority projects, totaling approximately 6,600 MW, is included in Appendix B.

At the time of this writing, the BLM had authorized or approved 5,200 MW of new RE capacity[10], with another 57,000 MW of applications pending for solar, wind and geothermal projects on BLM lands (see Table 8 and the Appendix). Throughout this paper, the term "pending" or "proposed" is used interchangeably to describe those projects where at the very least an application has been filed with BLM to develop a project on BLM land. There are many steps required between the time a developer files an application with BLM and when a project is finally approved and moves to the construction phase. The BLM only has responsibility for permitting; many other steps in the development process, such as financing, interconnection, and obtaining a power purchase agreement are completely outside of the BLM's control. The development status of these pending/proposed projects has not been tabulated for this paper; however, it is reasonable to assume that many of these projects will never be built.

Tables 8 and 9 indicate BLM's approved and pending projects by technology and by state. Approved or authorized projects total 5,236 MW[11] and expect to generate 11,697 GWh/year. The note below the tables indicate how the priority project lists from 2010-2012 are included in these numbers. Solar projects comprise more than 80% of the total capacity approved or authorized to date. No biopower or hydropower systems are proposed and no renewable energy projects have yet been proposed for BLM lands in Colorado or Montana.

---

[10] Since December 2009.
[11] Appendix A lists over 6,600 MW of approved projects. This is because an additional 1,400 MW was approved between the time Tables 8-10 were prepared and the time this publication went to print.

Table 8. Planned Renewable Nameplate Capacity by State on BLM Land

	Arizona	California	Colorado	Idaho	Montana	Nevada	New Mexico	Oregon	Utah	Washington	Wyoming	Total
Geothermal Approved (MW)	–	150	–	–	–	128	–	–	–	–	–	278
Solar Authorized (MW)	–	3,588	–	–	–	654	–	–	–	–	–	4,242
Wind Authorized (MW)	30	296	–	139	–	150	–	–	80	–	21	716
Total Approved or Authorized (MW)	30	4,034	–	139	–	932	–	–	80	–	21	5,236
Geothermal Pending (MW)	–	298	–	13	–	584	15	23	37	–	–	970
Solar Pending (MW)	18,308	11,618	–	–	–	16,437	2,200	–	–	–	–	48,563
Wind Pending (MW)	500	2,272	–	465	–	1,780	–	604	593	90	2,073	8,377
Total Pending (MW)	18,808	14,188	–	478	–	18,801	2,215	627	630	90	2,073	57,910
Total Approved and Pending (MW)	18,838	18,222	–	617	–	19,733	2,215	627	710	90	2,094	63,146

For Geothermal, pending includes 2011 and 2012 priority projects and only phase 3 proposed projects; approved includes 2010 priority projects and only phase 4 proposed projects. For solar and wind, pending includes 2011 and 2012 priority projects, authorized includes 2010 priority projects.

Source: BLM 2011

Table 9. Planned Renewable Generation by State on BLM Land

	Arizona	California	Colorado	Idaho	Montana	Nevada	New Mexico	Oregon	Utah	Washington	Wyoming	Total
Geothermal Approved (GWh)	–	1,117	–	–	–	953	–	–	–	–	–	2,070
Solar Approved (GWh)	–	6,286	–	–	–	1,146	–	–	–	–	–	7,432
Wind Approved (GWh)	92	908	–	426	–	460	–	–	245	–	64	2,195
Total Approved (GWh)	92	8,311	–	426	–	2,559	–	–	245	–	64	11,697
Geothermal Pending (GWh)	–	2,218	–	97	–	4,348	112	171	276	–	–	7,222
Solar Pending (GWh)	32,076	20,355	–	–	–	28,798	3,854	–	–	–	–	85,082
Wind Pending (GWh)	1,533	6,966	–	1,426	–	5,457	–	1,851	1,818	276	6,356	25,683
Total Pending (GWh)	33,609	29,539	–	1,522	–	38,604	3,966	2,022	2,094	276	6,356	117,987
Total Approved and Pending (GWh)	33,701	37,849	–	1,949	–	41,162	3,966	2,022	2,339	276	6,420	129,684

For Geothermal, pending includes 2011 and 2012 priority projects and only phase 3 proposed projects; approved includes 2010 priority projects and only phase 4 proposed projects. For solar and wind, pending includes 2011 and 2012 priority projects; authorized includes 2010 priority projects.

Source: BLM 2011

As shown in Table 10, the 5,200 MW of approved projects represents 11%-18% of the total capacity needed to meet RPS requirements in 2020.

Table 10. BLM RE Projects Potential Contribution to 2020 RPS Generation Requirements

		Arizona	California	Colorado	Idaho	Montana	Nevada	New Mexico	Oregon	Utah	Washington	Wyoming	Total
Renewable Energy Capacity (MW) Needed to Meet 2020 RPS Requirement (Table 5)	Minimum MW (High Capacity Factor, Low Capacity)	1,380	18,653	2,741	0	325	896	958	1,733	0	1,807	0	28,493
	Minimum MW (Low Capacity Factor, High Capacity)	2,144	31,354	2,741	0	662	896	1,421	3,079	0	3,238	0	45,535
Renewable Energy Capacity (MW) from Authorized BLM Projects		30	4,032	–	139	–	932	–	–	80	–	21	5,236
% of RPS Requirement from BLM Authorized Projects	Minimum MW (High Capacity Factor, Low Capacity)	2%	22%	0%	n/a	0%	104%	0%	0%	n/a	0%	n/a	18%
	Minimum MW (Low Capacity Factor, High Capacity)	1%	13%	0%	n/a	0%	104%	0%	0%	n/a	0%	n/a	11%

Source: BLM 2011

Geographic Locations of BLM Projects

All BLM priority (red circles) and proposed (yellow circles) RE projects are highlighted in Figure 9. BLM transmission projects are also highlighted in Figure 9. In addition, the figure shows the WECC foundational transmission lines, the five pilot transmission lines, the Western Governors' Association Western Renewable Energy Zones (WREZ) and the BLM solar energy zones (SEZ) as identified in the supplemental Solar PEIS. Additional details on the priority and proposed BLM solar, wind, geothermal and transmission projects are presented in the following sections.

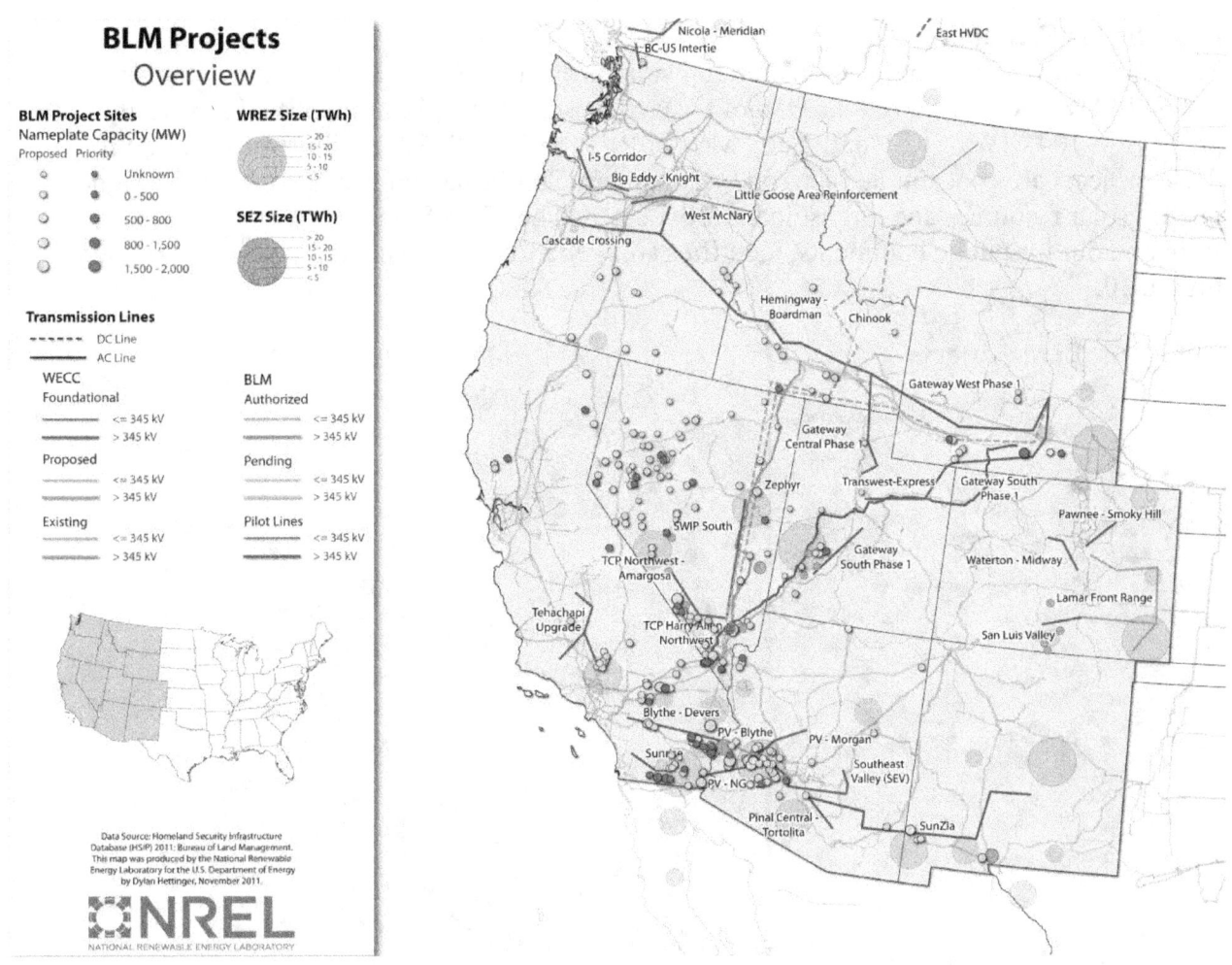

Figure 9. BLM projects proposed for the western United States.

BLM Solar Projects

The BLM has received applications for 104 solar projects identified for the WECC region (28,647 MW of CSP and 10,802 MW of PV); of these, 18 have been identified as priority projects (2,104 MW of CSP and 2,468 MW of PV) for 2010-2012. As illustrated in Figure 10, these projects are concentrated in Arizona, southern California and Nevada where the highest quality solar resources and largest loads are located. The BLM Solar PEIS, currently in draft form and out for public comments, identifies solar energy zones, which are also highlighted in Figure 10.

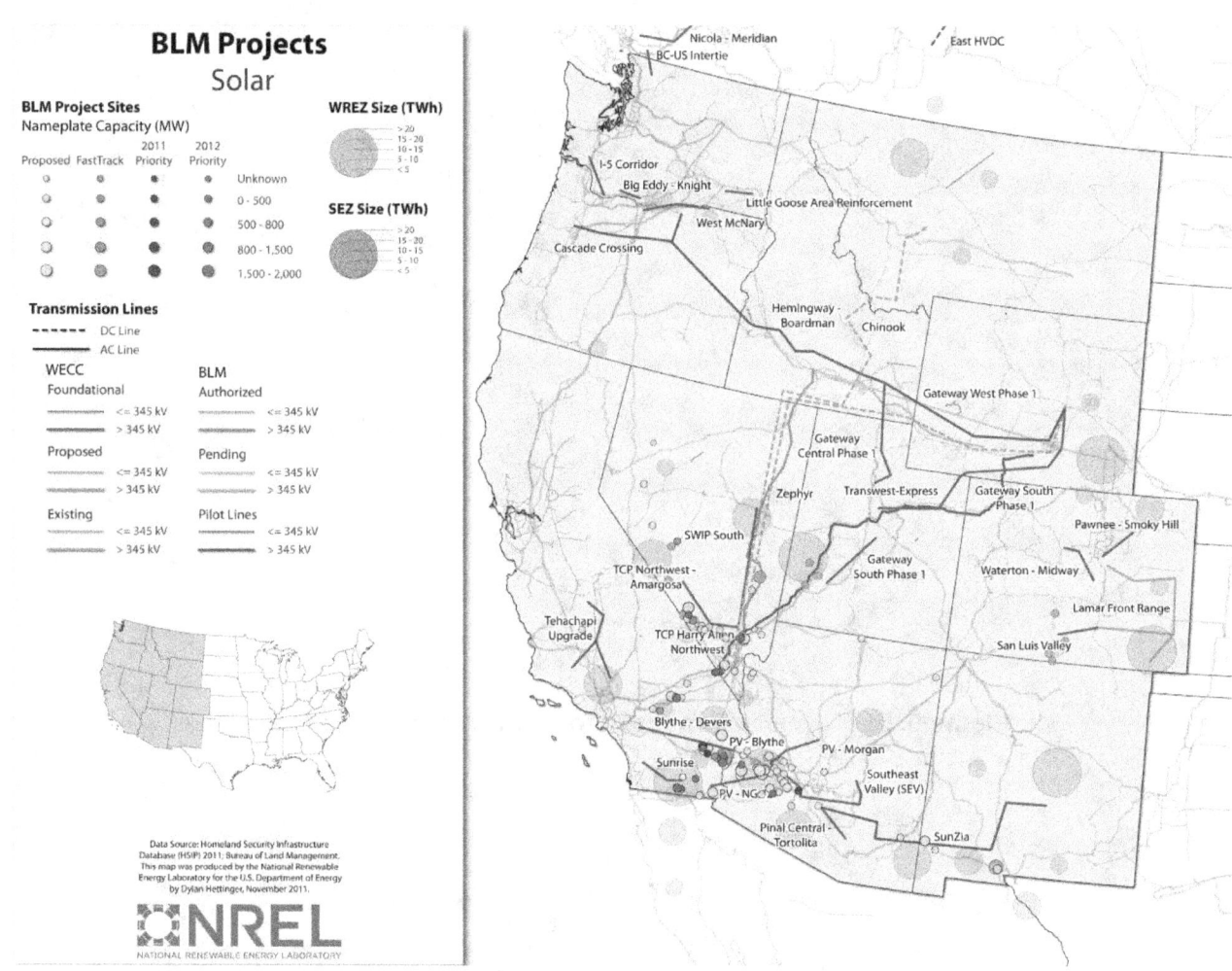

Figure 10. BLM solar projects proposed for the western United States.

BLM Wind Projects

The BLM has received applications for 76 wind projects in the WECC region (9,097 MW); of these, over 4,000 MW of projects have been identified as high priority projects (2010-2012). As illustrated in Figure 11, several of these projects are concentrated in southern California and along the northern border of California and Nevada; other wind projects are distributed throughout southern Oregon and Idaho, Arizona, Utah, and Wyoming where the highest quality wind resource is located.

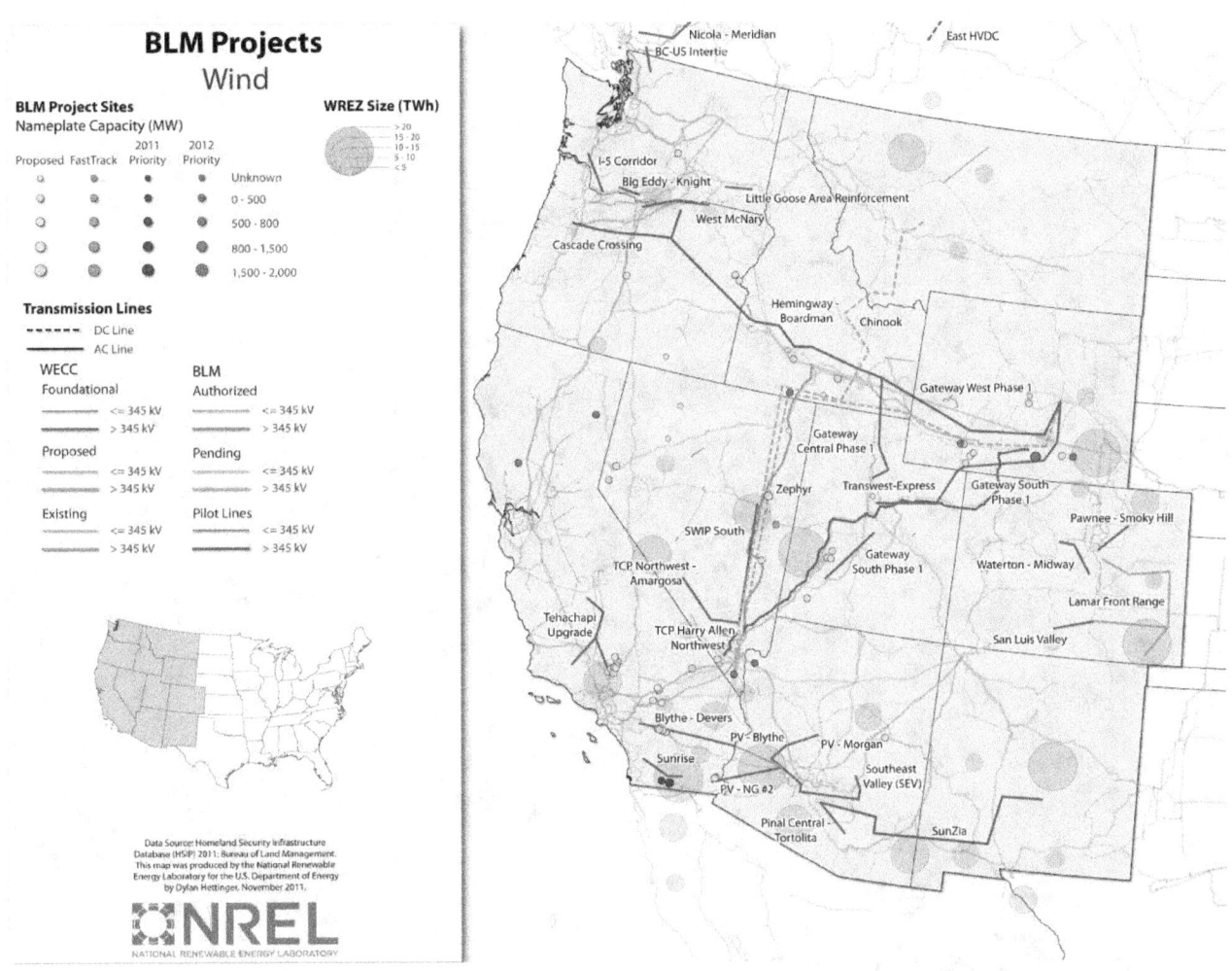

Figure 11. BLM wind projects proposed for the western United States.

BLM Geothermal Projects

The BLM has 109 geothermal projects identified for the WECC region; 51 projects are currently in the early stages of the approval process, 40 projects in the middle stage, and 14 projects totaling 1,431 MW are in late stages. Ten BLM geothermal projects have been identified as priority projects for 2010-2012 (523.5 MW). As illustrated in Figure 12, these projects are concentrated in Nevada where the highest quality geothermal resources are located. The Geothermal PEIS identified the BLM lands with potential for geothermal development that are also available for leasing; these areas are highlighted on Figure 12.

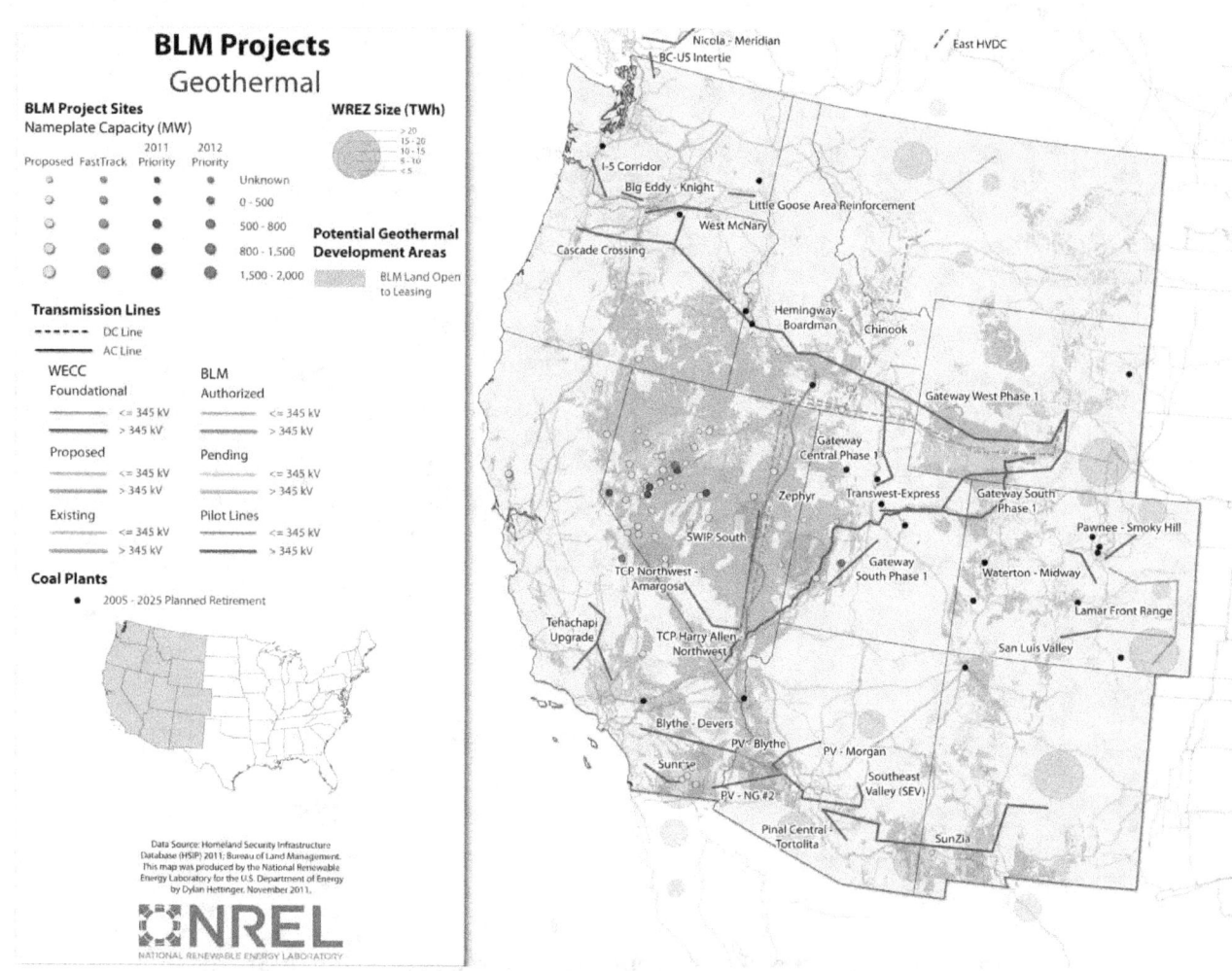

Figure 12. BLM geothermal projects proposed for the western United States.

Note: Only geothermal projects in the later stages of development are included in Figures 12 and 9 because the generation capacity of early-stage geothermal projects is generally not yet determined.

Coal retirements are based on either the official reported retirement date for existing plants, or if that is not available, on the assumption that coal units will retire in 65-75 years depending on size of the plant (Ventyx 2010).

Regional Transmission Planning Status

Increased investment in the nation's transmission infrastructure began several years ago in response to various needs including reliability and generator interconnection. As RE requirements increase, additional transmission infrastructure expansion will be required to connect high quality remote renewable resources with demand centers.

The cost effectiveness of a new transmission line depends on how much power it carries; lines carrying small amounts of power pose greater economic challenges. Small lines (230 kV or less) cost more per megawatt of carrying capability. A larger line costs less per megawatt, but that efficiency is lost if the line's capability is not fully utilized. Distance magnifies the effects of these factors, posing an extra economic challenge for small generation resources that are far from load.

In addition to economic challenges, environmental issues and other siting considerations often present barriers to the development of new transmission lines. Some of the key issues and considerations include potential impacts on individual animal and plant species and their habitats, cultural and historic resources, visual resources, and specially designated areas (e.g. parks, monuments, recreation drainages); land use and ownership; and the location of other infrastructure (e.g. pipelines, roads, and railways).

Several entities are planning, building, or tracking transmission projects. This report summarizes the status of BLM projects using SNL Financial tracking service information and WECC's recent study of transmission projects.

The BLM has 78 transmission projects in process, totaling 223 miles of less-than-230 kV lines, 996 miles of 230/345 kV lines, and 7,506 miles of lines over 500 kV. These lines are included in Figures 9-12. Existing transmission lines of 230 kV and 500 kV are also shown in Figures 9-12.

SNL Financial reports that a total of 19,577 miles of new transmission lines (between 115 kV and 500 kV) are currently planned for WECC states, with 879 miles currently under construction, 1,802 miles in the advanced development stage, and 16,896 miles announced (SNL Financial, 2011). The projects currently under construction and in advanced development are primarily intrastate and do not cross state lines. As with RE projects, not all the transmission projects that have been announced will be completed due to similar challenges.

Through its Regional Transmission Expansion Planning (RTEP) process, which is managed by the Transmission Expansion Planning Policy Committee, WECC is evaluating long term regional transmission needs in the region. The RTEP process seeks to include various factors that will impact long term transmission planning including demand, generation resources, policies, costs, reliability and emissions. WECC's RTEP process is meant to provide unbiased information and to advise and guide decision makers. WECC does not have authority to construct, site, permit, or fund transmission lines.

Under the RTEP process, WECC recently completed an initial study (WECC 2011) to identify the transmission projects that have a high probability of being in service in the next 10 years.

These projects, termed "foundational or common case projects" are included in each of the maps presented in this paper (WECC 2011). In total, these projects identified in the study would add more than 5,500 line miles to the existing 75,000 miles of existing transmission above 200 kV. According to the WECC study assumptions, renewable resources close to major load centers are largely utilized to fulfill the RPS in the 10-year planning studies. In the 2020 expected future network, every state with a renewable energy portfolio mandate or goal, other than Oregon and Utah, will receive more than 75% of their RPS energy from in-state resources. As more remote resources come online, RPS compliance beyond 2020 will likely require additional transmission (WECC 2011). In addition, if remote resources are selected in lieu of local generation for pending requirements, additional transmission will be required. Also, in some cases, long distance transmission to access remote renewable resources appears cost effective when compared to some local renewable generation. In the study, WECC also strongly recommended that states work together to implement regional transmission and resource development plans, as there may be economic and environmental benefits to such an approach.

Another transmission effort is developing around the newly formed Rapid Response Transmission Team and the recently announced "7 pilot transmission lines" (DOE 2011). The rapid response team seeks to coordinate and expedite the overall federal permitting, review and consultation processes for transmission lines. The seven pilot lines represent a concentrated effort by the federal government to expedite the permitting process of these stakeholder selected lines.

With the potential additions of the foundational projects, most of the major transmission paths do not appear to be overly congested in 2020. However, two major transmission paths, Montana to Northwest and Northwest to California, do appear to remain congested.

Western Renewable Energy Zones and EPAct Corridors
A number of studies have been done to identify the best locations for renewable energy projects in the WECC region from a renewable resource quality, transmission access, and low environmental impact perspective, including:

- Western Renewable Energy Zones Phase I Report (WECC 2009)
- EPAct Corridors PEIS.

The WREZs are the focus of regional transmission planning by the Western Governors' Association (WGA) and the Western Electricity Coordinating Council (WECC 2009) and are shown on each of the BLM project maps (Figures 9-12). The zones represent areas in the western region with high potential for large-scale development of renewable resources (where renewable resources are the most concentrated and are likely to have their highest capacity factors) and low environmental impacts. Potential resources that were quantified for each renewable energy zone (REZ) were screened not only for their native quality, but also for the likelihood that commercially significant development could occur in that area. The hub circles in the maps are scaled to show relative annual production potential of all renewable resources in a REZ (the circles do not indicate the perimeter of a development).[12]

The WREZ Phase I Report noted that the western United States contains a significant amount of commercially viable renewable energy resources outside of the potential WREZs. States are also developing state-specific REZs to meet RE needs at the least cost to state electricity customers; these states include Arizona, California, Colorado, Nevada, New Mexico, and Utah (these are not shown in the maps presented in this paper). The BLM has its own solar energy zones, called Solar Energy Development Areas, and is working to identify Wind Energy Priority Areas.

A number of comprehensive planning efforts to designate energy transport corridors across federal lands have been conducted with the goal of streamlining reviews and approvals of specific transmission projects crossing federal lands. Corridors are sited specifically to avoid as much as possible sensitive resources, land use conflicts, and extreme terrain while maximizing the opportunities to connect energy development areas with demand centers and support development of the existing transmission system.

Section 368 of EPAct requires the designation of "West-Wide" energy corridors on federal lands in the 11 WECC states and the establishment of procedures to ensure that additional corridors are identified and designated as necessary. Section 368 also requires processes to expedite applications to construct or modify oil, gas, and hydrogen pipelines and electricity transmission and distribution facilities. A number of federal agencies, including BLM, Bureau of Reclamation, U.S. Forest Service, and others proposed corridors for the transport and distribution of energy (electricity, oil, natural gas, and hydrogen) between supply and demand areas in the 11 western states while avoiding sensitive resources and land use and regulatory constraints to the fullest extent possible. If applicants develop energy transport projects within the proposed corridors, the resulting infrastructure would aid in alleviating congestion problems associated with electricity transmission in the western United States.

Corridors were sited using a four-step process that identified a number of important lands and resources to be avoided to the fullest extent possible. The agencies examined factors that constrain where a network of energy transport corridors could be located – including topographical, environmental, and regulatory constraints – as well as the overall suitability of particular lands to support development and operation of energy transport infrastructure. In some cases, corridors intersect or approach sensitive lands or resources. Most often these intersections follow existing infrastructure such as highways, transmission lines, or pipelines to avoid placing corridors in "greenfield" (undeveloped) locations.

Figure 13 shows the corridors designated as a result of EPAct Section 368. These corridors are not shown on the BLM RE project maps (Figures 9-12).

---

[12] Each gray-shaded circle indicates a REZ hub. A hub represents a conceptual step-up transformer where the electricity generated by all renewable resources in the REZ would get onto the transmission system. Hub circles are scaled to show relative annual production potential of all renewable resources in the REZ. Circles are not intended to indicate precise location of a new substation; actual collection point may be anywhere in the vicinity of the hub.

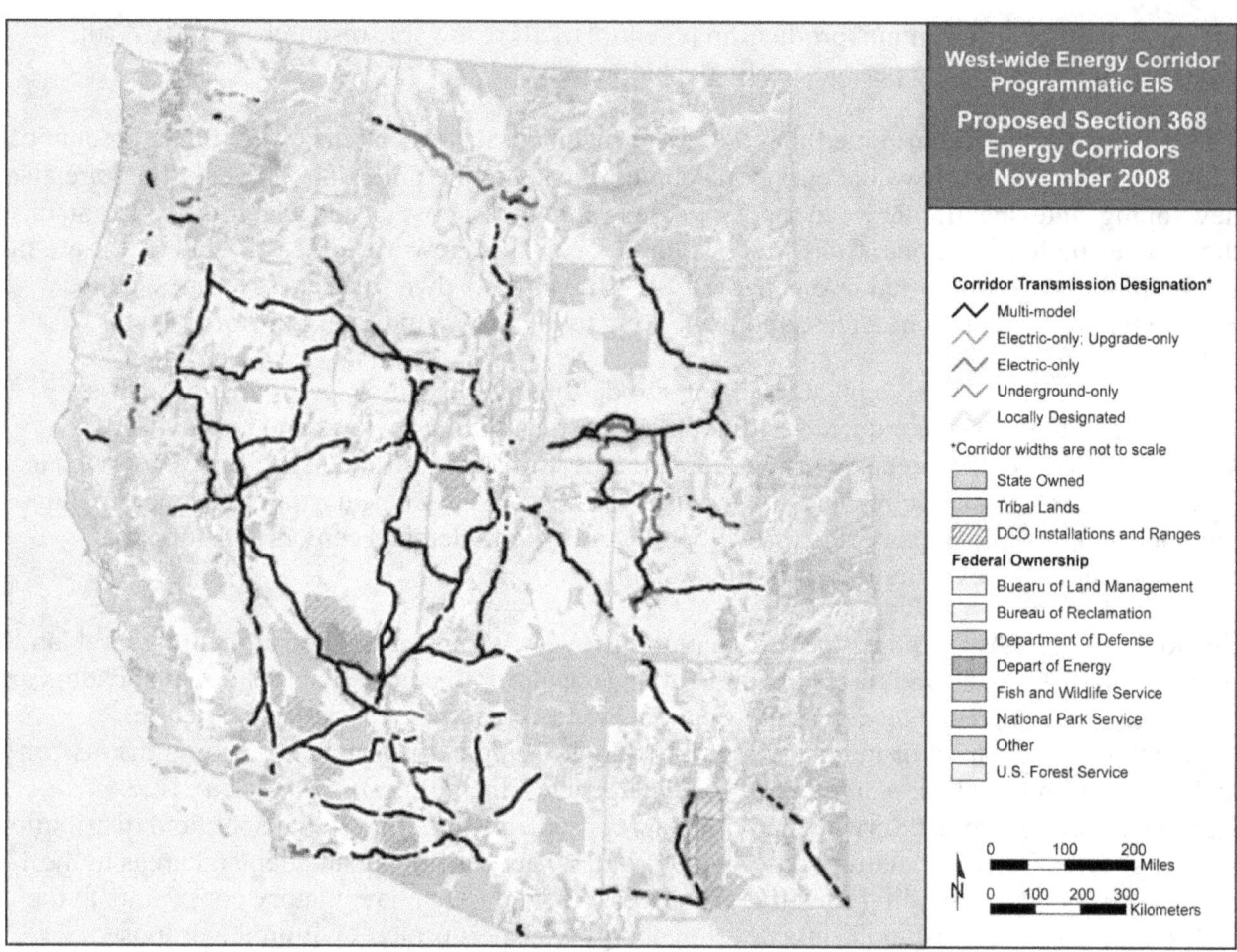

Figure 13. EPAct 368 corridors in the western states (Nov 2008).

Eighty-two percent of the energy corridors are located on BLM-managed lands, while 16% are on U.S. Forest Service lands. The remaining proposed corridor segments are on lands managed by DOI's Bureau of Reclamation and National Park Service, or by the Department of Defense. Individual projects proposed for these corridors would undergo further project-specific environmental analysis before being granted permits or rights of way.

Strategic Opportunities for BLM Renewable Energy Projects

Given the potential constraints in RPS markets, the BLM may want to look at factors beyond those currently considered in determining priority status to identify the best opportunities for proposed RE projects on federal lands. The BLM objectives of being "smart from the start" will take on new meaning over the next decade, and the following considerations may be helpful.

The BLM and DOI would likely benefit from identifying and permitting the highest quality projects on public lands because these will have the greatest chance of being built. Highest quality projects include those involving states which may fall short of their RPS requirements, and projects that are the lowest cost, nearest to existing transmission or load centers, or specific opportunities such as CSP projects located near fossil-fuel plants.

Alignment with Renewable Energy Zones

Figures 10-12 indicate that the BLM solar projects generally align with WREZs, while the BLM wind and geothermal projects generally do not. As highlighted in Figure 10, the BLM 2010 Fast Track and 2011 Priority Solar projects generally align with the WREZs. Most of the other proposed BLM solar projects generally align with the WREZs. Outliers are the pending projects in northern Arizona and central and southern Nevada that are relatively remote from the WREZs. As shown in Figure 11 the 2011 BLM wind priority projects generally align with the WREZ, but the 2012 BLM priority wind projects do not. In addition, many of the pending BLM wind projects do not align with WREZ. In particular, the BLM wind projects in Oregon, Idaho and Nevada are quite remote from the WREZ. As shown in Figure 12, four of the 10 BLM fast track and priority geothermal projects are well aligned with the WREZ. In general, the BLM pending geothermal projects are not aligned with the WREZ.

Alignment with Existing and Planned Transmission Expansion

The BLM planned projects within five miles of existing, WECC foundational or proposed, or BLM planned transmission are highlighted in Figure 14. These are the most promising BLM projects, from the perspective of proximity to existing and planned transmission lines. Overall, only one of the BLM priority projects, the Sierra Pacific Power Company Geothermal Project, in Douglas County, Nevada is not sited within five miles of an existing or WECC foundational transmission line. Most of the BLM proposed RE projects are within 20 miles of existing transmission lines and more than half are within five miles of existing transmission lines.

31

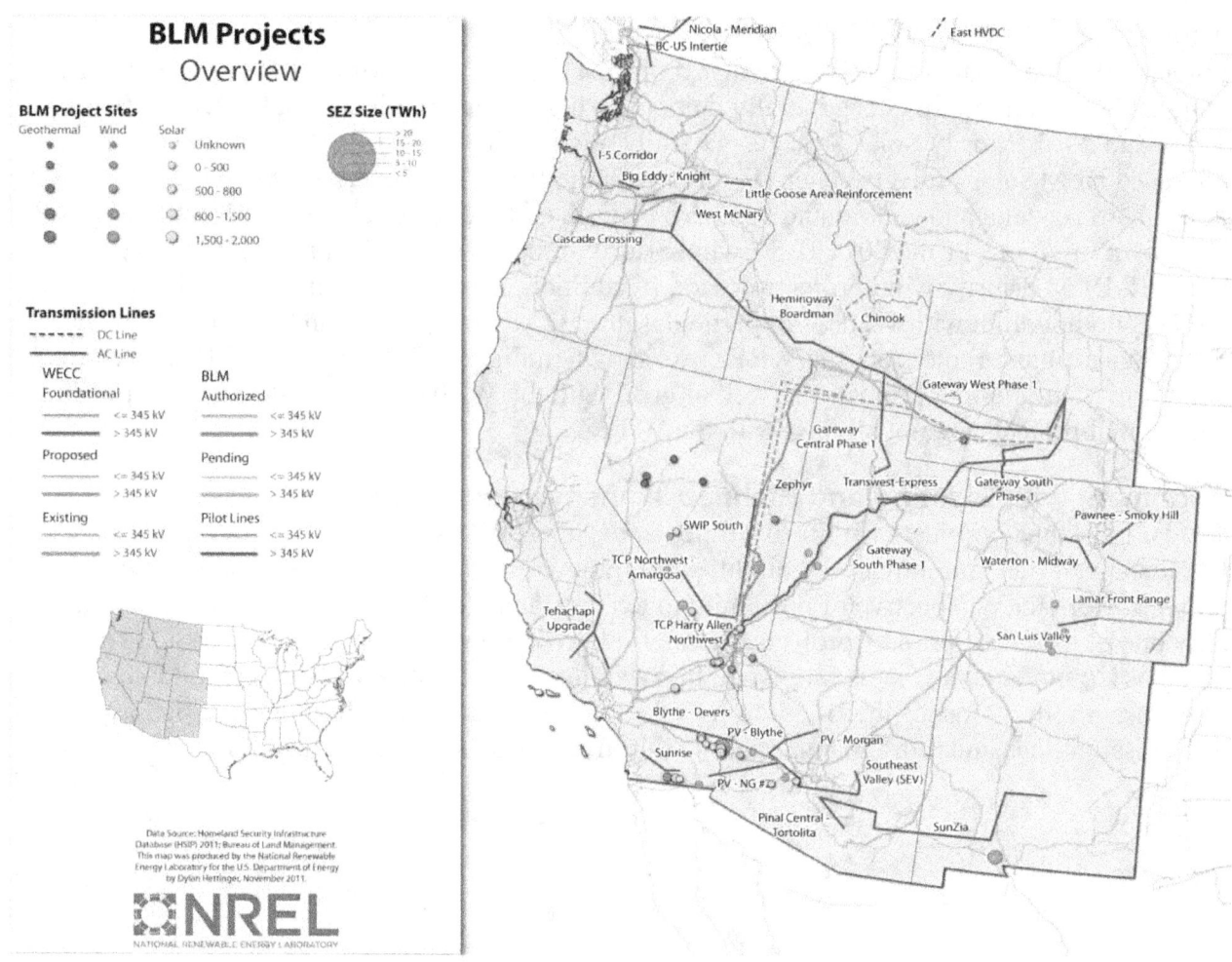

Figure 14. BLM projects within five miles of current, foundational, or BLM lines.

The *least* promising BLM RE planned projects, from the perspective of proximity to transmission lines, are summarized below for each technology.

Geothermal: All but two of the BLM priority geothermal projects are within 20 miles of existing transmission lines. Of these, the Fort Cove Enel Geothermal project in Millard County, Utah is within 20 miles of a WECC foundational transmission line project; however, the Fallon/Salt Wells project in Douglas County, Nev., is not within 20 miles of any existing or proposed transmission lines included in this report. The proposed BLM geothermal projects, in advanced stages of planning referred to as stages 3 and 4, with the most limited transmission access (beyond 20 miles from existing transmission) are listed in Table 11.

32

Table 11. BLM Proposed Geothermal Projects beyond 20 Miles of Transmission

Priority Project Year (2010/2011/2012)	2011	2012	
State	Nevada	Utah	Oregon
Serial Number	NVN-087795	UTU-085604	
Customer Name	Fallon / Salt Wells	Cove Fort	U.S. Geothermal
Project Name and/or Geographic Area	Sierra Pacific Power Co.	Enel Geothermal	Neal Hot Springs
Status (Pending Authorized)	Approved	Pending	Pending
Phase (Geothermal Only)			Phase 3
County	Douglas	Millard	Malheur
Within 5-miles of Existing	no	no	no
Within 20-miles of Existing	no	no	no
Within 20-miles of WECC Foundational	no		
Within 20-miles of WECC Proposed	no	no	no

Solar: All but two of the BLM priority solar projects are within five miles of existing transmission lines. These two California projects, the Genesis/Ford Dry Lake project in Riverside County and the Lucerne Valley project in San Bernardino County, are each within 20 miles of existing lines. The 17 proposed BLM solar projects with the most limited transmission access (beyond 20 miles from existing transmission) are listed in Table 12.

Table 12. BLM Proposed Solar Projects beyond 20 Miles of Transmission

Project Type	Solar								
State	Nevada	Nevada	Nevada	Nevada	Nevada	Nevada	Nevada	Nevada	Nevada
Serial Number	NVN-084359	NVN-089659	NVN-086571	NVN-084466	NVN-085657	NVN-087366	NVN-087756	NVN-089656	NVN-084465
Customer Name	Solar Millennium LLC	Element Power	Abengoa Solar Inc.	Pacific Solar Investments Inc.	Cogntrix Solar Services LLC	Solar Millennium LLC	Solar Millennium LLC	Element Power	Pacific Solar Investments Inc.
Project Name and/or Geographic Area	Amargosa Farm Road		Lathrop Wells Solar	Amargosa South	Amargosa South				Iberdrola, Amargosa North
County	Nye	Nye	Nye	Nye	Nye	Nye	Nye	Nye	Nye
Within 5-miles of Existing	no	no	no	no	no	no	no	no	no
Within 20-miles of Existing	no	no	no	no	no	no	no	no	no
Within 20-miles of WECC Foundational	no	no	no	no	no	no	no	no	no
Within 20-miles of WECC Proposed	no	no	no	no	no	no	no	no	no

Project Type	Solar (cont.)							
State	Nevada	Nevada	Nevada	Nevada	Nevada	Nevada	Nevada	Nevada
Serial Number	NVN-089560	NVN-084704	NVN-086782	NVN-086248	NVN-086249	NVN-086246	NVN-088156	NVN-086350
Customer Name	Gasna 39 LLC	Amargosa Flats Energy LLC	Southwest Solar Land Co. LLC	Ausra Nv I LLC	Ausra Nv I LLC	Ausra Nv I LLC	Ewindfarm Inc.	Solar Reserve LLC
Project Name and/or Geographic Area		Crystal / Johnnie	South Solar Ridge	Highway 160	Spector Range	Skeleton Hills / Lathrop Wells	Yucca Mountain Test Site	Pahroc Solar
County	Clark	Nye	Nye	Nye	Nye	Nye	Nye	Lincoln
Within 5-miles of Existing	no	no	no	no	no	no	no	no
Within 20-miles of Existing	no	no	no	no	no	no	no	no
Within 20-miles of WECC Foundational								
Within 20-miles of WECC Proposed	no	no	no	no	no	no	no	no

Source: BLM 2011a

Wind: All but one BLM priority wind project—the Sand Hills project in Albany County, Wyoming—is within 20 miles of existing transmission lines. The proposed BLM wind projects with the most limited transmission access (beyond 20 miles from existing transmission) are listed in Table 13.

Table 13. BLM Proposed Wind Projects beyond 20 Miles of Transmission

Project Type	Wind								
Priority Project Year (2010/2011/2012)						2012			
State	Idaho	Nevada	Oregon	Oregon	Oregon	Wyoming	California	Wyoming	Nevada
Serial Number	IDI-036923	NVN-088201	OROR-064395	OROR-065553	OROR-065616	WYW-166407	CACA-052186	WYW-142464	NVN-087411
Customer Name	City of Oakley	Nevada Wind Co.	Joseph Millworks Inc.	Horizon Wind Energy NWX	Horizon Wind Energy NWX	Shell Wind Energy Inc.	Renewable LLC	Pacific Corp.	Wilson Creek Wind Co.
Project Name and/or Geographic Area	n/a	Ely	Lime	Pueblo Mountain	Burnt River	Sand Hills	n/a	Foote Creek Rim I Wind	Wilson Creek
County	Cassia	White Pine	Baker	Harney	Baker	Albany	Imperial	Carbon	Lincoln
Within 5-miles of Existing	no	no	no	no	no	no	no	no	no
Within 20-miles of Existing	no	no	no	no	no	no	no	no	no
Within 20-miles of WECC Foundational	no	no	no	no	no	no			
Within 20-miles of WECC Proposed		no	no	no	no	no	no		

Source: BLM 2011a

Coal Plant Retirements

With the ability to provide baseload capacity to the grid, geothermal projects could replace generation capacity lost as conventional coal plants are retired. The coal plants expected to retire by 2012 are highlighted in Figure 12. Coal retirements are based on either the official reported retirement date for existing plants, or if that is not available, on the assumption that coal units will retire in 65-75 years depending on size of the plant (Ventyx 2010). For the most part, the coal plants slated for near-term retirement are not located within the BLM lands identified as areas with geothermal potential that could be leased. Two proposed projects near the Oregon-Idaho border are near two of these coal plants and could potentially back fill the generation lost by the retiring facilities. However, both of these projects (Weiser in Washington County, Idaho and Neal Hot Springs II in Malheur County, Oregon) are only in the first phase of the development process and thus would not likely be online quickly enough to take advantage of this opportunity. There may be additional opportunities for CSP with molten salt storage to play a role in replacing some of the baseload capacity associated with retiring coal plants. Also, with transmission expansion and the development of more sophisticated integration strategies and technologies, it is likely that there will be greater opportunities in the future to balance renewable resources with other renewable resources across the region.

Co-location of BLM Projects with Fossil Fuel Plants

Another area for maximizing the strategic value of BLM or other DOI projects would be to co-locate renewable technologies with existing or planned natural gas or coal-fired power plants. For example, the 2012 high priority solar project entitled Dry Lake Valley Solar (NV Energy) proposes to site a concentrating solar plant (CSP) next to an existing natural gas plant. An opportunity for a tribal project could involve development on the Navajo Indian Reservation, where preliminary analysis (Hurlbut 2011) indicates favorable economics for integrating CSP into the 2,200 MW Navajo Generating Station coal-fired power plant.

Conclusions

Absent new policy drivers and without the extension of the DOE loan guarantee program and Treasury's 1603 program, state RPS requirements are likely to remain a primary driver for new RE deployment in the western United States. Assuming no additional policy incentives are implemented, projected RE demand for the WECC states by 2020 is 134,000 GWh. Installed capacity to meet that demand will need to be within the range of 28,000-46,000 MW.

BLM projects that are currently authorized or approved should provide 11%-18% of the total capacity (9% of the generation) needed to meet the 2020 RPS requirements across WECC. If all of the currently authorized and 2011/2012 high priority BLM RE projects are deployed, RE projects on public lands will support 35% of the 2020 total requirement for new RE capacity.

With 5,200 MW of RE authorized or approved, and approximately 8,000 MW of additional 2011 and 2012 high priority projects, the BLM appears to be on track to meet the EPAct 2005 requirement of approving 10,000 MW of RE on public lands by 2012.

Most of the BLM's priority wind and geothermal projects are within 20 miles of existing transmission lines; most of BLM's priority solar projects are within five miles of existing transmission lines. New RE development projects sited close to load centers are not expected to be constrained by the current transmission infrastructure over the next 10 years; however, for more remotely sited RE projects, which some of the BLM projects will likely be, additional transmission infrastructure will be required. The additional 19,577 miles of new transmission lines (between 115 and 500 kV) currently planned for WECC states will support some of the expansion required for RE deployment required under western state mandatory RPS.

Based on the analyses of supply and demand in WECC states, and BLM's interest in leasing land for renewable energy projects, the following suggestions may be helpful for BLM:

Update the renewable energy project list. The information on BLM projects presented in this report changes frequently. NREL suggests that at least once per year, BLM go through its master project list and obtain updates on the status of the projects on its books. The BLM Washington office has recently issued a call for information and data for GIS analysis of the wind and solar projects that BLM's state and field offices are tracking. This information should be useful for updating that status of projects for FY12.

Focus on high-value project sites. The integration of BLM renewable energy projects with planned transmission lines (especially the five pilot lines) will take on greater significance over the next few years, and BLM lands within potential interconnection distance of these lines are likely to see increased interest by industry. Also, any BLM lands that are located close to load in the desert southwest in states that are potentially falling short on RPS requirements may see increased interest from developers. BLM should consider screening these lands, and the solar energy zones and other regions undergoing landscape-level planning, against criteria designed to identify prime sites for development of competitive leasing requests.

Work with other federal agencies and developers to facilitate project siting. A number of federal agencies, including other DOI bureaus and agencies, the Department of Defense,

Department of Homeland Security, Department of Agriculture, Department of Energy, and Department of Commerce are interested in deploying renewable energy technologies to meet their internal mission goals. As specific examples, the Environmental Protection Agency's RE-Powering America's Land program seeks to promote the development of renewable energy projects on brownfield sites such as abandoned mining lands, landfills, and contaminated lands. The Bureau of Indian Affairs and many tribes are working to develop renewables on tribal lands. Developers are also targeting state, local and private lands. Just as BLM has done with its Restoration Design project in Arizona, the BLM could benefit from continuing to work cooperatively with other agencies to identify the most suitable locations for development, regardless of land ownership.

Site projects to help support critical national needs. Similar to the strategy of siting projects to take advantage of projects and interests by other agencies and private developers, there may be opportunities to increase the strategic value of BLM's renewable energy projects by co-locating in areas that would support national or regional energy security and resiliency, and support national environmental goals. As an example of this, BLM projects could be sited in locations that, in an emergency situation, could help provide power supply for critical loads such as water pumping and treatment facilities, hospitals, military installations, National Guard facilities, critical substations, radar sites, data centers, and other high value loads. In some cases, BLM may choose to work with the developers and recommend a shift of the BLM projects to other locations in the region that may offer greater strategic advantages. Continuing to avoid projects on environmentally sensitive lands will support national environmental goals. GIS analysis can help identify these specific opportunities.

Identify options to integrate projects into existing fossil fuel generation. Siting renewable energy projects near old, retiring, or seldom used fossil fuel plants takes advantage of existing infrastructure and potential synergies. For example, BLM lands located near existing coal or gas plants may be candidate sites for solar thermal plants, including those with thermal storage, that are constructed from the outset to integrate fully into existing plants.

References

American Wind Energy Association (2011). *U.S. Wind Industry Year-End 2010 Market Report.* http://www.awea.org/learnabout/publications/upload/4Q10_market_outlook_public.pdf

Barbose, Galen (2010). Unpublished analysis accessed via personal correspondence.

Barbose, Galen (2011). National Renewable Portfolio Standards Summit. Oct 26-27, 2011. http://www.cleanenergystates.org/assets/Uploads/2011-RPS-Summit-Combined-Presentations-File.pdf . Source data used for this reference was received by personal correspondence November 2011 and used for graphs and calculations in this report.

Bird, Lori; Hurlbut, David; Donohoo, Pearl; Cory, Karlynn; Kreycik, Claire (2010). *An Examination of the Regional Supply and Demand Balance for Renewable Electricity in the United States through 2015: Projecting from 2009 through 2015.* NREL Report No. TP-6A2-45041. Golden, CO: National Renewable Energy Laboratory, 52 pp.

Energy Policy Institute (January 2011). "Transmission Siting and Public Lands: Options for Improvement and the West Case Study." http://epi.boisestate.edu/media/5831/epi_transmission%20siting%20and%20public%20lands%20january%202011_final.pdf

Energy Policy Act (EPACT) of 2005 (2005). http://doi.net/iepa/EnergyPolicyActof2005.pdf, accessed on November 27, 2011

Geothermal Energy Association (2011). *Annual U.S. Geothermal Power Production and Development Report.* http://geo-energy.org/pdf/reports/April2011AnnualUSGeothermalPowerProductionandDevelopmentReport.pdf

Heeter, Jenny; Bird, Lori (2011). *Status and Trends in U.S. Compliance and Voluntary Renewable Energy Certificate Markets (2010 Data).* NREL Report No. TP-6A20-52925. *Golden, CO:* National Renewable Energy Laboratory, 66 pp.

Hurlbut, David (2011). *Navajo Generating Station and Air Visibility Regulations: Alternatives and Impacts* (Forthcoming). Golden, CO: National Renewable Energy Laboratory.

Reuters (2010). *Utility Cancels Contract for Tessera's Calico Solar Project,* 12/27/10, http://blogs.reuters.com/environment/2010/12/28/utility-cancels-contract-for-tesseras-calico-solar-project/. Accessed December 1, 2011.

Sherwood, Larry (2011). *US Solar Market Trends 2010.* http://irecusa.org/wp-content/uploads/2011/06/IREC-Solar-Market-Trends-Report-June-2011-web.pdf

Solar Energy Industries Association and GTM Research (2011). *U.S. Solar Market Insight 2010 Year-In-Review.*

SNL Financial (2011a). http://www.snl.com/SNLWebPlatform/Content/Home/Home.aspx (accessed 11/11/11). Source data downloaded using SNL's Excel Add-in tool.

http://www.snl.com/Interactivex/snlxl.aspx. Accessed 11/11/11. Data set: Power Plants, Units In Development (Under Construction, Planned).

SNL Financial (2011b). Transmission Projects. http://www.snl.com/interactivex/TransProjects.aspx. Accessed July 7, 2011.

SNL Financial (2011c). http://www.snl.com/interactivex/TransProjects.aspx. Accessed 12/14/11.

Turchi, Craig; Langle, Nicholas; Bedilion, Robin; Libby, Cara (2011). *Solar Augment Potential of U.S. Fossil-Fired Power Plants.* NREL/TP-5500-50597. Golden, CO: National Renewable Energy Laboratory, 25 pp.

U.S. Council on Environmental Quality, Rapid Response Transmission Team. http://www.whitehouse.gov/administration/eop/ceq/initiatives/interagency-rapid-response-team-for-transmission. Accessed November 30, 2011.

U.S. Department of Energy (2011). *Obama Administration Announces Job-Creating Grid Modernization Pilot Projects.* http://energy.gov/articles/obama-administration-announces-job-creating-grid-modernization-pilot-projects. Accessed November 30, 2011.

U.S. Department of the Interior/Bureau of Land Management (2005). Wind Energy Development Programmatic EIS. http://windeis.anl.gov/eis/what/index.cfm

U.S. Department of the Interior/Bureau of Land Management (2008). Geothermal Resources Leasing Programmatic EIS. http://www.blm.gov/wo/st/en/prog/energy/geothermal/geothermal_nationwide.html

U.S. Department of the Interior (2009). Renewable Energy Development by the Department of the Interior. http://elips.doi.gov/app_so/act_getfiles.cfm?order_number=3285. Accessed December 23, 2011.

U.S. Department of the Interior/Bureau of Land Management (2009). Renewable Energy Coordination Offices. http://www.blm.gov/pgdata/etc/medialib/blm/wo/Communications_Directorate/public_affairs/news_release_attachments.Par.48600.File.dat/09SecOrderRenewableEnergyOfc0116.pdf. Accessed November 27, 2011.

U.S. Department of the Interior/Bureau of Land Management (2010). Draft Solar Energy Development Programmatic EIS. http://solareis.anl.gov/index.cfm

U.S. Department of the Interior/Bureau of Land Management (2010a). Public Land Statistics. http://www.blm.gov/public_land_statistics/pls10/pls10_combined.pdf

U.S. Department of the Interior/Bureau of Land Management (2010b). *Renewable Energy and the BLM Factsheet.* http://www.blm.gov/pgdata/etc/medialib/blm/wo/MINERALS__REALTY__AND_RESOURCE

40

PROTECTION/energy/renewable_references.Par.95879.File.dat/2010%20Renewable%20Ene
rgy%20headed.pdf

U.S. Department of the Interior/Bureau of Land Management (2011a). Unpublished data
accessed via personal correspondence.

U.S. Department of the Interior/Bureau of Land Management (2011b). RE Priority Projects.
http://www.blm.gov/priorityprojects.

U.S. Energy Information Administration (2010a). *Table 5. Residential Average Monthly Bill by
Census Division and State.* http://www.eia.gov/cneaf/electricity/esr/table5.html. Accessed July 6,
2011.

U.S. Energy Information Administration (2010b). *Form EIA-861 Final Data File for 2009.*
http://www.eia.gov/cneaf/electricity/page/eia861.html. Accessed May 1, 2011.

U.S. Energy Information Administration (2010c). *Annual Energy Outlook 2010.*
http://www.eia.gov/oiaf/archive/aeo10/pdf/0383%282010%29.pdf. Accessed July 11, 2011.

U.S. Energy Information Administration (2011a). *Form EIA-906, EIA-920, and EIA-923 Data.
2010: EIA-923 January - December, Excel Format.*
http://www.eia.gov/cneaf/electricity/page/eia906_920.html. Accessed July 6, 2011.

U.S. Energy Information Administration (2011b). *Form EIA-860 Data Files.*
http://www.eia.gov/cneaf/electricity/page/eia860.html. Accessed July 6, 2011.

U.S. Energy Information Administration (2011c). *Table 10: All States, All Sectors.*
http://www.eia.gov/electricity/data.cfm#sales. Accessed November 21, 2011.

U.S. Energy Information Administration (2011d). *Net Generation by State by Type of Producer
by Energy Source, Annual Back to 1990 (EIA-906, EIA-920, and EIA-923).*
http://www.eia.gov/electricity/data.cfm#generation. Accessed November 21, 2011.

U.S. Energy Information Administration (2011e). Annual Electric Utility Data File – EIA-861.
http://205.254.135.24/cneaf/electricity/page/eia861.html. Accessed November 28, 2011.

Ventyx Velocity Suite, ©2010. Accessed via http://www.ventyx.com/.

Western Electricity Coordinating Council (2009). *WREZ Phase 1 Report (2009).*
http://www.westgov.org/rtep/219.

Western Electricity Coordinating Council (2011). *10-Year Regional Transmission Plan: Plan
Summary.* http://www.wecc.biz/library/StudyReport/Documents/Plan_Summary.pdf

Xcel Energy (2011). *Resource Plan for Energy Needs through 2018.*
http://www.xcelenergy.com/About_Us/Energy_News/News_Releases/Xcel_Energy_files_2011_
resource_plan_for_energy_needs_through_2018. Accessed December 1, 2011.

Appendix A.

List of Renewable Energy Projects Approved by BLM Between June 2009 and December 2011 (6,587 MW)

SOLAR (Total MW = 5,637; Total Peak Jobs = 10,631)					
#	State	Project Name (Developer)	Capacity (MW)	Peak Jobs	Status
1	CA	Ivanpah° (BrightSource Energy)	370 MW (power tower)	1,100	Under Construction
2	NV	Silver State North (First Solar)	50 MW (PV)	310	Under Construction
3	CA	Genesis° (NextEra)	250 MW (parabolic)	1,135	Under Construction
4	NV	Crescent Dunes° (Solar Reserve)	110 MW (power tower)	550	Under Construction
5	CA	Mohave Solar*° (Abengoa)	250 MW (parabolic)	890	Under Construction
6	CA	Desert Sunlight° (First Solar)	550 MW (PV)	645	Under Construction
7	CA	Lucerne Valley (Chevron Energy Solutions/Fotowatio)	45 MW (PV)	48	Authorized
8	NV	Amargosa Farm Road (Solar Millennium)	464 MW (parabolic)	1,370	Authorized
9	CA	Imperial Solar Energy Center South* (CSOLAR)	200 MW (PV)	255	Authorized
10	CA	Imperial Solar Energy Center West* (CSOLAR)	250 MW (PV or CPV)	290	Authorized
11	CA	Imperial Valley (acquired by AES Solar)	709 MW (change to PV)	900	Authorized^Δ
12	CA	Calico (acquired by K Road Power)	663.5 MW (PV/solar dish)	536	Authorized^Δ
13	CA	Blythe (Solar Millennium)	1,000 MW (change to PV)	1,361	Authorized^Δ
14	AZ	Sonoran Solar (NextEra)	300 MW (solar trough)	374	Authorized
15	CA	Rice Solar Energy* (Rice Solar Energy LLC)	150 MW (power tower)	500	Authorized
16	CA	Centinela Solar Energy (Centinela Solar Energy, LLC)	275 MW (thin film PV)	367	Authorized
WIND (Total MW = 544; Total Peak Jobs = 1,224)					
#	State	Project Name (Developer)	Capacity (MW)	Peak Jobs	Status
1	NV	Spring Valley Wind (Pattern Energy)	150 MW	237	Under Construction
2	CA	West Butte Wind* (West Butte Wind Power)	104 MW	415	Authorized
3	CA	Tule Wind (Iberdrola)	186 MW	337	Authorized
4	OR	Echanis Wind/North Steens Transmission*	104 MW	235	Authorized
GEOTHERMAL (Total MW = 406; Total Peak Jobs = 700)					
#	State	Project Name (Developer)	Capacity (MW)	Peak Jobs	Status
1	NV	Blue Mountain Geothermal Power Plant	49 MW	175	In Operation
2	NV	Jersey Valley* (Ormat)	30 MW	70	In Operation
3	NV	McGinness Hills* (Ormat)	90 MW	153	Under Construction
4	NV	Hot Sulfur Springs/Tuscarora*° (Ormat)	15 MW	25	Under Construction
5	NV	Coyote Canyon (Terragen)	62 MW	105	Authorized
6	NV	Salt Wells (Ormat)	40 MW	68	Authorized
7	NV	Salt Wells (Vulcan)	120 MW	104	Authorized

* Project is on private lands but requires BLM connected action for approval of transmission or facilities.
° Project has received a DOE "1705" Loan Guarantee.
^Δ The company has decided to change technology, so the NEPA for these projects must be updated.

Appendix B. BLM's 2012 High Priority Project List (as of 12/21/11)

SOLAR (Total MW = 2,414 (1,464 on BLM land)					
#	State	Project Name (Developer)	Capacity (MW)	Current NEPA Status	Target Decision Date
1	AZ	Quartizite Solar Energy (Solar Reserve)	100	On-going	Target 2012 Q3
2	CA	McCoy Solar (NextEra)	750	NOI 8/29/2011	Target 2012 Q4
3	CA	Desert Harvest (enXco)	100	NOI Pending	Target 2012 Q4
4	CA	Calico Solar Redesign (K Power)	[664]	NOI 10/25/2011	Target 2012 Q4
5	CA	Ocotillo Sol (San Diego Gas & Electric)	14	NOI 7/17/2011	Target 2012 Q3
6	CA	Mount Signal Solar Farm #1* (82LV 8ME, LLC)	600 (private)	EA Pending	Target 2012 Q3
7	NV	Amargosa North Solar (Pacific Solar Investments)	150	Draft EIS Pending	Target 2012 Q4
8	NV	Silver State South (First Solar)	350	NOI	Target 2012 Q4
9	NV	Moapa Solar* (K Road Moapa Solar)	350 (Tribal)	Draft EIS 11/25/2011	Target 2012 Q3

WIND (Total MW = 4,180 (2,382 on BLM land)					
#	State	Project Name (Developer)	Capacity (MW)	Current NEPA Status	Status
1	AZ	Mohave County Wind Farm (BP Wind)	500	Draft EIS Pending	Target 2012 Q4
2	CA	Walker Ridge Wind (AltaGas)	70	Draft EIS Pending	Target 2012 Q4
3	NV	Searchlight Wind (Duke Energy)	20	Draft EIS Pending	Target 2012 Q4
4	WY	Sand Hills Ranch (Shell Wind Energy)	50 total (4 BLM)	Final EA Pending	Target 2012 Q3
5	WY	Choke Cherry/Sierra Madre (Power Company of WY)	3000 total (1500 BLM)	Final EIS Pending	Target 2012 Q3
6	WY	White Mountain (Teton Wind LLC)	360 total (108 BLM)	EA FONSI 8/15/2011	Target 2012 Q3

GEOTHERMAL (Total MW - 95)					
#	State	Project Name (Developer)	Capacity (MW)	Current NEPA Status	Status
1	CA	Casa Diablo (Mammoth Pacific)	33	NOI Published	Target 2012 Q4
2	NV	New York Canyon (Terragen)	62	EA Pending	Target 2012 Q3

* Connected Action, [ ] = not counted in total

www.ingramcontent.com/pod-product-compliance
Lightning Source LLC
Chambersburg PA
CBHW081243180526
45171CB00005B/525